WMRK

Lecture Notes in Physics

New Series m: Monographs

Springer

Berlin
Heidelberg
New York
Barcelona
Budapest
Hong Kong
London
Milan
Paris
Santa Clara
Singapore
Tokyo

The Editorial Policy for Monographs

The series Lecture Notes in Physics reports new developments in physical research and teaching - quickly, informally, and at a high level. The type of material considered for publication in the New Series m includes monographs presenting original research or new angles in a classical field. The timeliness of a manuscript is more important than its form, which may be preliminary or tentative. Manuscripts should be reasonably self-contained. They will often present not only results of the author(s) but also related work by other people and will provide sufficient motivation, examples, and applications.

The manuscripts or a detailed description thereof should be submitted either to one of the series editors or to the managing editor. The proposal is then carefully refereed. A final decision concerning publication can often only be made on the basis of the complete manuscript, but otherwise the editors will try to make a preliminary decision as definite as they can on the basis of the available information.

Manuscripts should be no less than 100 and preferably no more than 400 pages in length. Final manuscripts should preferably be in English, or possibly in French or German. They should include a table of contents and an informative introduction accessible also to readers not particularly familiar with the topic treated. Authors are free to use the material in other publications. However, if extensive use is made elsewhere, the publisher should be informed. Authors receive jointly 50 complimentary copies of their book. They are entitled to purchase further copies of their book at a reduced rate. As a rule no reprints of individual contributions can be supplied. No royalty is paid on Lecture Notes in Physics volumes. Commitment to publish is made by letter of interest rather than by signing a formal contract. Springer-Verlag secures the copyright for each volume.

The Production Process

The books are hardbound, and quality paper appropriate to the needs of the author(s) is used. Publication time is about ten weeks. More than twenty years of experience guarantee authors the best possible service. To reach the goal of rapid publication at a low price the technique of photographic reproduction from a camera-ready manuscript was chosen. This process shifts the main responsibility for the technical quality considerably from the publisher to the author. We therefore urge all authors to observe very carefully our guidelines for the preparation of camera-ready manuscripts, which we will supply on request. This applies especially to the quality of figures and halftones submitted for publication. Figures should be submitted as originals or glossy prints, as very often Xerox copies are not suitable for reproduction. For the same reason, any writing within figures should not be smaller than 2.5 mm. It might be useful to look at some of the volumes already published or, especially if some atypical text is planned, to write to the Physics Editorial Department of Springer-Verlag direct. This avoids mistakes and time-consuming correspondence during the production period.

As a special service, we offer free of charge LATEX and TEX macro packages to format the text according to Springer-Verlag's quality requirements. We strongly recommend authors to make use of this offer, as the result will be a book of considerably improved technical quality.

Manuscripts not meeting the technical standard of the series will have to be returned for improvement.

For further information please contact Springer-Verlag, Physics Editorial Department II, Tiergartenstrasse 17, D-69121 Heidelberg, Germany.

Gerald Dunne

Self-Dual
Chern-Simons Theories

Springer

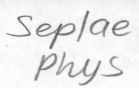

Author

Gerald Dunne
Physics Department
University of Connecticut
Storrs, CT 06269, USA

Cataloging-in-Publication Data applied for

Die Deutsche Bibliothek - CIP-Einheitsaufnahme

Dunne, Gerald:
Self-dual Chern-Simons theories / Gerald Dunne. - Berlin ;
Heidelberg ; New York ; Barcelona ; Budapest ; Hong Kong ;
London ; Milan ; Paris ; Tokyo : Springer, 1995
  (Lecture notes in physics : N.s. M, Monographs ; 36)
  ISBN 3-540-60257-7
NE: Lecture notes in physics / M

ISBN 3-540-60257-7 Springer-Verlag Berlin Heidelberg New York

© Springer-Verlag Berlin Heidelberg 1995
Printed in Germany

Typesetting: Camera-ready by authors using T$_E$X
SPIN: 10481135   55/3142-543210 - Printed on acid-free paper

For Elyse

Preface

This book is intended to provide a pedagogical introduction to the subject of self-dual Chern-Simons theories, which comprises a relatively recent addition to the 'zoo' of self-dual field theories. I have tried to give a detailed presentation of the basic features and properties of these theories. It is not possible to cover the entire subject in an introductory text, so I have treated some of the more advanced developments in less detail; nevertheless, I have attempted to balance this compromise with an extensive bibliography and an annotated overview of more recent and current work.

This book grew from various seminars and lectures I have given over recent years. I thank Jihn E. Kim and Choonkyu Lee for the opportunity to present some of this material at the XIIIth *Symposium on Theoretical Physics (Field Theory and Mathematical Physics)* held at Mt. Sorak (Korea) in 1994. Some material is also drawn from a series of seminars at the University of Connecticut.

I have learned a great deal from the seminars and research papers of colleagues working in this field, and I would like especially to acknowledge the pioneering contributions of Roman Jackiw, Choonkyu Lee, Kimyeong Lee, So-Young Pi and Erick Weinberg.

It was a pleasure to collaborate with Roman Jackiw, So-Young Pi and Carlo Trugenberger on some of this material. In addition, I am grateful for discussions and correspondence with Giovanni Amelino-Camelia, Dongsu Bak, Daniel Cangemi, Andrea Cappelli, Dan Freedman, Kurt Haller, Peter Horvathy, Bum-Hoon Lee, Alberto Lerda and Guillermo Zemba.

Special thanks are due to Bob Jaffe for his generous advice and assistance concerning the publication of this book.

I am grateful for partial support from the Department of Energy and from the University of Connecticut Research Foundation.

VII

Contents

X

I. INTRODUCTION

A. Self-Dual Theories

"Self-duality" is a powerful notion in classical mechanics and classical field theory, in quantum mechanics and quantum field theory. It refers to theories in which the interactions have particular forms and special strengths such that the second order equations of motion (in general, a set of coupled nonlinear partial differential equations) reduce to first order equations which are simpler to analyze. The "self-dual point", at which the interactions and coupling strengths take their special self-dual values, corresponds to the minimization of some functional, often the energy or the action. This gives self-dual theories crucial *physical* significance. For example, the self-dual Yang-Mills equations have minimum action solutions known as instantons, the Bogomol'nyi equations of self-dual Yang-Mills-Higgs theory have minimum energy solutions known as 't Hooft-Polyakov monopoles, and the planar Abelian Higgs model has minimum energy self-dual solutions known as Nielsen-Olesen vortices. Instantons, monopoles and vortices have become paradigms of topological structures in field theory and quantum mechanics, with important applications in particle physics, astrophysics, condensed matter physics and mathematics. In these Lecture Notes, I discuss a new class of self-dual theories, *self-dual Chern-Simons theories*, which involve charged scalar fields minimally coupled to gauge fields whose 'dynamics' is provided by a Chern-Simons term in $2 + 1$ dimensions (*i.e.* two spatial dimensions). The physical context in which these self-dual Chern-Simons models arise is that of anyonic quantum field theory, with direct applications to such planar models as the quantum Hall effect, anyonic superconductivity and Aharonov-Bohm scattering. In addition, there are further interesting connec-

1

tions with the more mathematically inspired theory of integrable models.

A novel feature of these self-dual Chern-Simons theories is that they permit a realization with *either* relativistic *or* nonrelativistic dynamics for the scalar fields. Self-dual theories for instantons, monopoles and (Nielsen-Olesen) vortices arise from relativistic field theories. In the nonrelativistic case, the self-dual point of the Chern-Simons system corresponds to a quartic scalar potential, with overall strength determined by the Chern-Simons coupling strength. The nonrelativistic self-dual Chern-Simons equations may be solved completely for all finite charge solutions, and the solutions exhibit many interesting relations to two dimensional (Euclidean) integrable models. The classification of these finite charge solutions is also strongly analogous to the classification of instanton solutions in four dimensional Euclidean spacetime. In the relativistic case, while the general exact solutions to the self-duality equations are not explicitly known, the solutions correspond to topological and nontopological solitons and vortices, many characteristics of which can be deduced from algebraic and asymptotic data. These self-dual Chern-Simons theories also have the property that, at the self-dual point, they may be embedded into a model with an extended supersymmetry. This serves as another explicit illustration of this general feature of self-dual theories - indeed, recent work has sought to *characterize* self-duality directly in terms of a related extended supersymmetry structure. This approach is especially attractive, as it provides a bridge between the classical and the quantum analysis of these models.

Before introducing the self-dual Chern-Simons theories, in this Introduction I briefly review some other important self-dual theories, in part as a means of illustrating the general idea of self-duality, but also because various specific properties of these theories will appear

2

explicitly in our analysis of the self-dual Chern-Simons systems.

Perhaps the most familiar, and in a certain sense the most fundamental, self-dual theory is that of four dimensional self-dual Yang-Mills theory. The Yang-Mills action is

$$\mathcal{S}_{YM} = \int d^4x \ tr \ (F_{\mu\nu} F_{\mu\nu}) \tag{1.1}$$

where $F_{\mu\nu} = \partial_\mu A_\nu - \partial_\nu A_\mu + [A_\mu, A_\nu]$ is the gauge field curvature. The Euler-Lagrange equations form a complicated set of coupled nonlinear partial differential equations:

$$D_\mu F_{\mu\nu} = 0 \tag{1.2}$$

where $D_\mu = \partial_\mu + [A_\mu, \ ]$ is the covariant derivative. However, in four dimensional Euclidean space the Yang-Mills action (1.1) is minimized by solutions of the self-dual (or anti-self-dual) Yang-Mills equations:

$$F_{\mu\nu} = \pm \tilde{F}_{\mu\nu} \tag{1.3}$$

where $\tilde{F}_{\mu\nu} \equiv \frac{1}{2}\epsilon_{\mu\nu\rho\sigma} F_{\rho\sigma}$ is the dual field strength. Note that the self-duality equations (1.3) are first order equations (in contrast to the second order equations of motion (1.2)), and their "instanton" solutions are known in detail (for a review see [211,212,168]). Also, solutions to the self-duality equations (1.3) are automatically solutions to the original Euler-Lagrange equations (1.2) since the dual field strength $\tilde{F}_{\mu\nu}$ satisfies the Bianchi identity $D_\mu \tilde{F}_{\mu\nu} = 0$, independent of any equations of motion. We shall see that the nonrelativistic self-dual Chern-Simons equations have many interesting connections with these self-dual Yang-Mills equations.

Another important class of self-dual equations consists of the "Bogomol'nyi equations"

$$D_i \Phi = -\epsilon_{ijk} F_{jk} \tag{1.4}$$

where Φ is a scalar field. The Bogomol'nyi equations arise in the theory of magnetic monopoles in $3 + 1$ dimensional space-time. They come from a minimization of the static energy functional of a Yang-Mills-Higgs system in a special parametric limit known as the Bogomol'nyi-Prasad-Sommerfield (BPS) limit [210,24]. It is interesting to note that these Bogomol'nyi equations can also be obtained from the (anti-) self-dual Yang-Mills equations (1.3) by a 'dimensional reduction' from four dimensions to three dimensions. In the anti-self-dual Yang-Mills equations (1.3) consider all fields to be independent of x^4, and identify A_4 with the scalar field Φ. Then the equations (1.3) reduce to

$$F_{41} = F_{23} \qquad\qquad \rightarrow \qquad\qquad D_1\Phi = -F_{23}$$

$$F_{42} = -F_{13} \qquad\qquad \rightarrow \qquad\qquad D_2\Phi = F_{13}$$

$$F_{43} = F_{12} \qquad\qquad \rightarrow \qquad\qquad D_3\Phi = -F_{12} \qquad (1.5)$$

which are precisely the Bogomol'nyi equations (1.4). We shall see that the nonrelativistic self-dual Chern-Simons equations may also be obtained from the self-dual Yang-Mills equations by a similar dimensional reduction, although in that case the dimensional reduction is from four dimensions to two dimensions. Furthermore, the *relativistic* self-dual Chern-Simons equations involve a special algebraic embedding problem (that of embedding $SU(2)$ into the gauge algebra) which also plays a crucial role in the analysis of the Bogomol'nyi equations (1.4).

The abelian Higgs model in $2 + 1$ dimensions is a model of a complex scalar field ϕ interacting with a $U(1)$ gauge field with conventional Maxwell dynamics. This model supports vortex solutions known as Nielsen-Olesen vortices, and is a relativistic analogue of

4

the Landau-Ginzburg phenomenological model for superconductivity [196,24,137]. The static energy functional for this system is

$$E = \int d^2x \left[\frac{1}{2}B^2 + \left|\vec{D}\phi\right|^2 + V\left(|\phi|\right) \right] \tag{1.6}$$

This static energy may be re-expressed as

$$E = \int d^2x \Big[|(D_j - i\epsilon_{jk}D_k)\,\phi|^2 + \frac{1}{2}\left(B + |\phi|^2 - v^2\right)^2 + v^2 B$$

$$-\frac{1}{2}\left(|\phi|^2 - v^2\right)^2 + V\left(|\phi|\right) \Big] \tag{1.7}$$

Thus, for a special "self-dual" quartic potential,

$$V_{\text{self-dual}} = \frac{1}{2}\left(|\phi|^2 - v^2\right)^2 \tag{1.8}$$

the static energy functional is bounded below by (v^2 times) the magnetic flux, and this "Bogomol'nyi bound" is saturated by solutions to the following set of self-duality equations:

$$D_j\phi = i\,\epsilon_{jk}D_k\phi \tag{1.9a}$$

$$F_{12} = v^2 - |\phi|^2 \tag{1.9b}$$

The term "self-duality" arises from the appearance of the duality operation in (1.9a). These equations have vortex solutions which have been studied in great detail [196,24,137,245]. We shall see that the self-duality equations in the self-dual Chern-Simons systems also arise from minimizing the energy functional in a $2+1$ dimensional theory, and the resulting Chern-Simons self-duality equations have a very similar form to the abelian Higgs model self-duality equations (1.9). This is true of both the relativistic and the nonrelativistic self-dual Chern-Simons systems.

Yang [252] proposed an approach to the four dimensional self-dual Yang-Mills equations (1.3) in which they can be viewed as the consistency conditions for a set of first order differential operators. This idea is fundamental to the notion of "integrability" of systems of differential equations, a subject with many deep connections to self-dual theories [241–243,255,42,114]. Indeed, it has been conjectured that all low-dimensional integrable systems can be obtained from the self-dual Yang-Mills equations (1.3) by some sort of dimensional reduction, for some particular gauge algebra. This relationship has been explicitly studied for classic integrable systems such as the nonlinear Schrödinger equation, the KdV equation, the Sine-Gordon equation and others. The general relation is due to the zero-curvature approach to integrability, which has a natural formulation in terms of gauge theory [206,82,42].

If the self-dual Yang-Mills equations (1.3) are rewritten in terms of the null coordinates $u = (x^1 + ix^2)/\sqrt{2}$ and $v = (x^3 + ix^4)/\sqrt{2}$, they become

$$F_{uv} = 0$$

$$F_{\bar{u}\bar{v}} = 0$$

$$F_{u\bar{u}} + F_{v\bar{v}} = 0 \qquad (1.10)$$

These equations express the consistency conditions for the first order equations

$$(D_u - \zeta D_{\bar{v}})\,\psi = 0$$

$$(D_v + \zeta D_{\bar{u}})\,\psi = 0 \qquad (1.11)$$

where ζ is known as a "spectral parameter". The first two equations in (1.10) can be solved locally to give

$$A_u = H^{-1}\partial_u H \qquad\qquad A_v = H^{-1}\partial_v H$$

$$A_{\bar{u}} = K^{-1}\partial_{\bar{u}}K \qquad\qquad A_{\bar{v}} = K^{-1}\partial_{\bar{v}}K \qquad (1.12)$$

where H and K are gauge group elements. Then, defining $J = HK^{-1}$, the third of the self-duality equations in (1.10) take the 'Yang' form [252]

$$\partial_{\bar{u}}\left(J^{-1}\partial_u J\right) + \partial_{\bar{v}}\left(J^{-1}\partial_v J\right) = 0 \qquad (1.13)$$

If we now make a dimensional reduction in which the fields are chosen to be independent of x^2 and x^4, this equation becomes the two dimensional equation

$$\partial_\mu\left(J^{-1}\partial_\mu J\right) = 0 \qquad (1.14)$$

which is known as the chiral model equation. The chiral model equation will play a very important role in our analysis of the nonrelativistic self-dual Chern-Simons equations. Also note that if $J \in SU(N)$ and J is further restricted to satisfy the condition $J^2 = \mathbf{1}$, then (1.14) is the equation of motion for the CP^{N-1} model, which is yet another well-known self-dual system [211,256].

The final class of models that we recall in this Introduction are known as Toda theories. The original Toda system described the displacements of a line of masses joined by springs with an *exponential* spring tension [230]. This provides a beautiful nonlinear, but still integrable, generalization of the standard linear (Hooke's law) model. The equations of motion for the Toda lattice are

$$\ddot{y}_i = -C_{ij}e^{y_j} \qquad (1.15)$$

where the matrix C_{ij} is the tridiagonal discrete approximation to the second derivative, and can be chosen for periodic or open boundary conditions. This system is classically integrable in the limit of an infinite number of masses, in the sense that it possesses an infinite number of conserved quantities in involution. The Toda lattice system also has a deep algebraic structure due to the fact that the matrix C_{ij} in (1.15) is the Cartan matrix of the Lie algebra $SU(N)$ (or its affine extension). Indeed, this relationship allows one to extend the original Toda system to a Toda lattice based on other Lie algebras.

The Toda system generalizes still further, to an integrable set of nonlinear *partial* differential equations in *two* dimensions

$$\nabla^2 y_i = -C_{ij}e^{y_j} \qquad (1.16)$$

where ∇^2 is the two dimensional Laplacian. These Toda equations are not only integrable, but also *solvable*, in the sense that the solution may be written in terms of $2r$ arbitrary functions, where r is the rank of the classical Lie algebra whose Cartan matrix appears in (1.16) [158,177,192]. For $SU(2)$ the classical Toda system reduces to the nonlinear Liouville equation

$$\nabla^2 y = -2e^y \qquad (1.17)$$

which was solved by Liouville [182], and which has played a significant role in string theory and models of quantum gravity. Both the Liouville and Toda equations, together with their solutions, appear prominently in the analysis of the nonrelativistic self-dual Chern-Simons models. Moreover, the Toda equations also arise from the Bogomol'nyi equations (1.4) when one looks for spherically symmetric monopole solutions [179]. This reduction involves an algebraic embedding problem very similar to one that appears in the treatment of the relativistic self-dual Chern-Simons models.

B. Chern–Simons Theories: Basics

The self-dual Chern-Simons theories discussed in these Lecture Notes describe charged scalar fields in $2 + 1$ dimensional space-time, minimally coupled to a gauge field whose dynamics is given by a Chern-Simons Lagrangian rather than by a conventional Maxwell (or Yang-Mills) Lagrangian. The possibility of describing gauge theories with a Chern-Simons term rather than with a Yang-Mills term is a special feature of odd-dimensional space-time, and the $2 + 1$ dimensional case is especially distinguished in the sense that the derivative part of the Chern-Simons Lagrangian is *quadratic* in the gauge fields. To conclude this Introduction, I briefly review some of the important properties [46,249,54,55] of the Chern-Simons Lagrange density:

$$\mathcal{L}_{CS} = \epsilon^{\mu\nu\rho} tr \left(\partial_\mu A_\nu A_\rho + \frac{2}{3} A_\mu A_\nu A_\rho \right) \tag{1.18}$$

The gauge field A_μ takes values in a finite dimensional representation of the gauge Lie algebra $\mathcal{G}$. The totally antisymmetric ϵ-symbol $\epsilon^{\mu\nu\rho}$ is normalized with $\epsilon^{012} = 1$. In an abelian theory, the gauge fields A_μ commute, and so the trilinear term in (1.18) vanishes due to the antisymmetry of the ϵ-symbol. The Euler-Lagrange equations of motion derived from this Lagrange density are simply

$$F_{\mu\nu} = 0 \tag{1.19}$$

which follows directly from the fact that

$$\frac{\delta \mathcal{L}_{CS}}{\delta A_\mu} = \epsilon^{\mu\nu\rho} F_{\nu\rho} \tag{1.20}$$

At first sight, the equations of motion (1.19) seem somewhat trivial, with solutions that are simply pure gauges $A_\mu = g^{-1} \partial_\mu g$. However, much recent work has revealed that with the inclusion of topological effects and external sources these equations are far from trivial, as they possess an extraordinarily rich structure with important applications to conformal field theory, quantum groups and condensed matter physics [249,25,76,257,90].

For our purposes here, it is very important to notice that the equations of motion (1.19) are *first-order* in space-time derivatives, in contrast to the Yang-Mills equations of motion (1.2) which are second-order. One viewpoint of self-dual theories is that the equations of motion are such that they may be factorized into simpler first-order equations. But for a gauge theory with a Chern-Simons Lagrange density, the gauge equation of motion is *already* first-order and so it can be used directly as one of the self-duality equations. Later, we shall see in detail how this fact proves useful in the context of the self-dual Chern-Simons models.

The equations of motion (1.19) are gauge covariant under the gauge transformation

$$A_\mu \to A_\mu^g \equiv g^{-1} A_\mu g + g^{-1} \partial_\mu g \tag{1.21}$$

and so the Lagrange density (1.18) defines a sensible gauge theory (at least at the classical level) even though the Lagrange density (1.18) itself is not invariant under the gauge transformation (1.21). Indeed, under a gauge transformation $\mathcal{L}_{CS}$ transforms as

$$\mathcal{L}_{CS}(A) \to \mathcal{L}_{CS}(A) - \epsilon^{\mu\nu\rho} \partial_\mu tr \left(\partial_\nu g \, g^{-1} A_\rho \right)$$

$$- \frac{1}{3} \epsilon^{\mu\nu\rho} tr \left(g^{-1} \partial_\mu g g^{-1} \partial_\nu g g^{-1} \partial_\rho g \right) \tag{1.22}$$

For an abelian Chern-Simons theory, the final term in (1.22) vanishes and the change in $\mathcal{L}_{CS}$ is a total space-time derivative. Hence the *action*, $S = \int d^3x \mathcal{L}_{CS}$, is gauge invariant and so we expect that a sensible quantum gauge theory may be formulated. However, for a *non-abelian* Chern-Simons theory the final term in (1.22) is proportional

10

to the winding number of the group element g, and so the Chern-Simons action changes by a constant under a gauge transformation with nontrivial winding number. This has important implications for the development of a quantum nonabelian Chern-Simons theory. To ensure that the quantum amplitude $exp(iS)$ remains invariant, the Chern-Simons Lagrange density (1.18) must be multiplied by a dimensionless coupling parameter κ' which assumes quantized values [46,249]

$$\kappa' = \frac{integer}{4\pi}, \tag{1.23}$$

with standard normalizations. This argument for a quantized coupling parameter is reminiscent of Dirac's quantization condition for the quantum mechanical magnetic monopole [50,51,100], which was developed further for a field theory of monopoles by Schwinger [220–222]. In a field theory, which in general will require renormalization, one can ask whether the quantization condition (1.23) applies to the bare coupling or to the renormalized coupling. Fortunately, Chern-Simons theories permit a deeper probing of this question, with the result that at one-loop the bare coupling parameter in a nonabelian theory receives a finite additive renormalization shift which is such that $4\pi\kappa$ is shifted by an integer. Moreover, there are strong indications (both computational and fundamental) that this result is valid to all orders. These computations were initially performed for gauge theories involving both a Yang-Mills and a Chern-Simons term in the gauge field Lagrangian [205], and then extended to pure Chern-Simons theories [249,25,76]. More recently, this issue has been explored, with similar conclusions, with the inclusion of matter fields and also spontaneous symmetry breaking [33].

Another important feature of Chern-Simons theories is that the Chern-Simons term describes a *topological* gauge field theory [249,23] in the sense that there is no explicit dependence on the space-time

11

metric. This follows because the Lagrange density (1.18) can be written directly as a 3-form : $\mathcal{L}_{CS} = tr(AdA + A^3)$. Thus the action is independent of the space-time metric, and so the Chern-Simons Lagrange density $\mathcal{L}_{CS}$ does not contribute to the energy momentum tensor. This may also be understood by noting that $\mathcal{L}_{CS}$ is first order in space-time derivatives

$$\mathcal{L}_{CS} = \epsilon^{ij} tr\left(A_i \dot{A}_j\right) + tr\left(A_0 F_{12}\right) \tag{1.24}$$

The time derivative part of $\mathcal{L}_{CS}$ indicates the canonical structure of the theory, with A_1 and A_2 being canonically conjugate fields. This is radically different from conventional (Yang-Mills) gauge theory in which the gauge field components A_i may be regarded as coordinate fields, canonically conjugate to the electric field $E_i \equiv F_{0i}$. The A_0 part of the Lagrange density produces the Gauss law constraint, and there is no contribution to the Hamiltonian. This implies that the Chern-Simons gauge field does not have any real dynamics of its own - it is a nonpropagating field whose dynamics comes from the fields to which it is minimally coupled. The precise implementation of this classical canonical structure in a quantum theory leads to many interesting features in Chern-Simons theories.

The Chern-Simons Lagrange density (1.18) may be coupled to an external matter current J^μ as

$$\mathcal{L} = \frac{\kappa}{2}\mathcal{L}_{CS} - tr\left(A_\mu J^\mu\right) \tag{1.25}$$

(Note that we include the factor of $1/2$ in the Chern-Simons coupling coefficient for later convenience.) This leads to the equations of motion [generalizing Equation (1.19)]

$$F_{\mu\nu} = -\frac{1}{\kappa}\epsilon_{\mu\nu\rho}J^\rho \tag{1.26}$$

involving the covariantly conserved $(D_\mu J^\mu = 0)$ current. Thus, the time component J^0 of the matter current is proportional to the magnetic field

12

$$J^0 = \kappa F_{12} \tag{1.27}$$

which is the Chern-Simons Gauss law constraint, and which is important for the interpretation of Chern-Simons theories as field theories for anyons [175]. The spatial components J^i of the current are everywhere perpendicular to the electric field

$$J^i = \kappa \epsilon^{ij} F_{j0} \tag{1.28}$$

With this coupling to external matter, the equations of motion (1.26) are still first order in space-time derivatives acting on the fields (in contrast to the corresponding equations of motion, $D_\mu F^{\mu\nu} = J^\nu$, in conventional Yang-Mills theory). This fact is crucial for the formulation of self-dual theories involving Chern-Simons gauge fields. The relations (1.27-1.28) are fundamental to the application of Chern-Simons theories to condensed matter systems such as the quantum Hall effect [209,229,86,257].

The last property of Chern-Simons theories that we mention in this Introduction is that the Higgs mechanism behaves very differently when the gauge fields are Chern-Simons fields. In a conventional gauge theory, the Higgs mechanism produces a massive gauge mode in a broken vacuum in which the scalar field, to which the gauge fields are coupled, possesses a nonvanishing vacuum expectation value. Formally, this mechanism is independent of the dimension of spacetime, but in $2+1$ dimensions the possibility of including a Chern-Simons term for the gauge fields leads to a richer variety of mass generation effects. For example, even without any symmetry breaking at all, a gauge theory with both a Yang-Mills and a Chern-Simons term describes a massive dynamical gauge mode, with mass determined by the Chern-Simons coupling parameter κ, and with spin ± 1 given by the sign of κ. This system has been dubbed "topologically massive gauge theory" [46]. The novel mass and spin properties of fields in $2 + 1$ dimensions may be understood in terms of the representations of the corresponding Poincaré and Lorentz algebras [22,223,18,234]. Now if such a gauge field is coupled to a scalar field with a symmetry-breaking minimum in its potential, the 'Higgs mechanism' leads now to *two* massive modes, as one new mode is generated by combining the Goldstone boson with the longitudinal part of the gauge field in the standard manner, and the mass of the existing (in the unbroken vacuum) topologically massive photon is also shifted [205,246,110]. Furthermore, if the Yang-Mills term is not present (and so there is no massive gauge mode in the unbroken vacuum) then the 'Chern-Simons-Higgs mechanism' produces a *single* massive gauge excitation in a symmetry breaking vacuum [48]. However, since the Chern-Simons term is first-order in space-time derivatives, the mass of this gauge excitation is proportional to the *square* of the vacuum expectation value of the scalar field, in contrast to the conventional Higgs mechanism for which the gauge mass is proportional to the *magnitude* of the scalar field vacuum expectation value. The relativistic self-dual Chern-Simons theories described in these Lecture Notes involve a self-dual scalar field potential which possesses nontrivial symmetry breaking vacua. The quantum analysis of these models therefore provides an interesting forum for application of the Chern-Simons-Higgs mechanism. We shall see that the particular self-dual form of the scalar potential leads to intricate mass spectra in the broken vacua.

There is, of course, much more that could be written about Chern-Simons models, but the above brief outline covers the features that are most immediately relevant to the topics discussed in these Lecture Notes. Further details will be introduced as necessary at the appropriate points in the discussion throughout the book. For general field theoretical properties of Chern-Simons theories, the interested reader is referred to [46,249,123,208,11], with applications to conformal field theory discussed in [76,25]. For applications of Chern-Simons theories to anyon physics see [85,90,97,118,175,247]. General applications to planar condensed matter physics are discussed in [86,234]; with the quantum Hall effect in [209,229,257] and anyonic superconductivity in [185].

II. ABELIAN NONRELATIVISTIC MODEL

In this Chapter we introduce the abelian nonrelativistic self-dual Chern-Simons model [125]. This model serves as a field theoretical description of anyons - point particles in the plane with fractional statistics. In this Chapter we concentrate on the symmetry and self-dual structure of the classical field theory, deferring discussion of quantum aspects to Chapter 6. This $2 + 1$ dimensional theory has a Bogomol'nyi style lower bound for the energy when the scalar potential takes a purely quartic form, the overall strength of which depends on the Chern-Simons coupling κ. This energy lower bound (which is in fact zero) is saturated by solutions to a set of self-duality equations and, as a consequence of a special dynamical symmetry, these self-dual solutions exhaust all static solutions. The self-duality equations may be converted into the Liouville equation, an integrable nonlinear partial differential equation whose solutions are explicitly known. This therefore leads to a complete characterization of the vortex-like self-dual solutions. Other more complicated solutions may then be generated from these by suitable coordinate transformations. Finally, we show how to generalize this model to include Maxwell dynamics for the gauge field, while still preserving the self-dual structure.

A. Lagrangian Formulation

The abelian nonrelativistic self-dual Chern-Simons system describes a complex scalar field $\psi(\vec{x}, t)$ which is minimally coupled to an abelian gauge field $A_\mu(\vec{x}, t)$. This model is defined in two dimensional space (which is taken to be R^2, unless otherwise specified) and the dynamical equations for the matter field ψ (and its conjugate,

16

ψ^*) are nonrelativistic. However, the gauge field A_μ is most conveniently expressed using a "Lorentz covariant" notation $A_\mu = (A_0, \vec{A})$ with "Minkowski metric" $g_{\mu\nu} = diag(-1, 1, 1)$ and with "c" set to unity. The gauge field is chosen to have only a Chern-Simons term in the Lagrange density. This has the consequence that the gauge field does not possess any dynamics of its own - in fact, the gauge field is determined by the nonrelativistic matter. Thus, even though some of the formulae involving the gauge field *look* relativistic, this system is indeed nonrelativistic. The Lagrange density for this system is

$$\mathcal{L} = \frac{\kappa}{2}\epsilon^{\mu\nu\rho}A_\mu\partial_\nu A_\rho + i\psi^* D_0\psi - \frac{1}{2m}\left|\vec{D}\psi\right|^2 + \frac{g}{2}|\psi|^4 \qquad (2.1)$$

Here m is the mass of the scalar field ψ, κ is a coupling constant which determines the strength of the Chern-Simons term

$$\mathcal{L}_{\text{CS}} = \epsilon^{\mu\nu\rho}A_\mu\partial_\nu A_\rho, \qquad (2.2)$$

and g is a coupling constant which determines the strength of the $|\psi|^4$ nonlinearity. The totally antisymmetric symbol $\epsilon^{\mu\nu\rho}$ is chosen with $\epsilon^{012} = +1$. The gauge covariant derivative D_μ is defined as

$$D_\mu = \partial_\mu + iA_\mu \qquad (2.3)$$

The Euler-Lagrange equations of motion which follow from the Lagrange density (2.1) are

$$iD_0\psi = -\frac{1}{2m}\vec{D}^2\psi - g|\psi|^2\psi \qquad (2.4a)$$

$$F_{\mu\nu} = -\frac{1}{\kappa}\epsilon_{\mu\nu\rho}J^\rho \qquad (2.4b)$$

where $J^\mu \equiv (\rho, \vec{J})$ is a Lorentz covariant notation for the (conserved) nonrelativistic charge and current densities:

$$\rho = |\psi|^2 \qquad (2.5a)$$

17

$$J^j = -\frac{i}{2m}\left(\psi^* D^j \psi - \left(D^j \psi\right)^* \psi\right) \tag{2.5b}$$

The matter equation of motion (2.4a), together with the field-current relation (2.4b), is referred to as the *planar gauged nonlinear Schrödinger equation* [125].

The equations of motion (2.4) are invariant under the abelian gauge transformation

$$\psi \to e^{-i\lambda}\psi \qquad\qquad A_\mu \to A_\mu + \partial_\mu \lambda \tag{2.6}$$

The Lagrange density (2.1) is *not* invariant under this gauge transformation as the Chern-Simons term changes by a total derivative:

$$\mathcal{L}_{CS} \to \mathcal{L}_{CS} + \partial_\mu \left(\lambda \epsilon^{\mu\nu\rho}\partial_\nu A_\rho\right) \tag{2.7}$$

Nevertheless, the Lagrange density (2.1) defines a sensible gauge theory (at least at the classical level) because the equations of motion (2.4) are gauge invariant.

The gauge field equation of motion (2.4b) may be re-expressed as

$$B = \frac{1}{\kappa}\rho \tag{2.8a}$$

$$E^i = -\frac{1}{\kappa}\epsilon^{ij}J_j \tag{2.8b}$$

where the planar "magnetic field" strength is

$$B = \partial_1 A_2 - \partial_2 A_1 \equiv F_{12} \tag{2.9}$$

the planar "electric field" strength is

$$E^i = \partial_i A_0 - \partial_0 A_i \equiv F_{i0} \tag{2.10}$$

and we use the spatial antisymmetric symbol ϵ^{ij} with $\epsilon^{12} = +1$. In terms of these gauge invariant quantities, the equations of motion

18

(2.8) imply that the charge density ρ is proportional to the magnetic field B, and the current density $\vec{J}$ is perpendicular to the electric field $\vec{E}$. This fact is important for phenomenological applications of Chern-Simons theories as effective field theories for the quantum Hall effect [257]. Indeed, the abelian model discussed in this Chapter has been generalized to a multi-abelian $[U(1)]^N$ theory [149] (see also [73]), which is relevant for the *fractional* quantum Hall effect.

As a consequence of the Euler-Lagrange equations of motion (2.4), the charge and current densities satisfy the continuity equation

$$\dot{\rho} + \vec{\nabla} \cdot \vec{J} = 0 \tag{2.11}$$

which is conveniently expressed as

$$\partial_\mu J^\mu = 0 \tag{2.12}$$

using the "Lorentz covariant" notation. The net charge

$$Q \equiv \int d^2x \, \rho \tag{2.13}$$

is conserved and, as a consequence of (2.8a), is proportional to the total magnetic flux

$$\Phi \equiv \int d^2x \, B \tag{2.14}$$

with correspondence

$$\Phi = \frac{1}{\kappa} Q \tag{2.15}$$

B. Hamiltonian Formulation

The abelian nonrelativistic system described in Section II A may also be described in a Hamiltonian formulation, a fact that will prove useful in understanding the static solutions of the system. The Lagrange density (2.1) may be rewritten as

$$\mathcal{L} = i\psi^*\dot{\psi} + \kappa\dot{A}_1 A_2 - \frac{1}{2m}\left|\vec{D}\psi\right|^2 + \frac{g}{2}|\psi|^4 + A_0\left(\kappa B - \rho\right) \qquad (2.16)$$

where irrelevant total derivative terms have been dropped. From this first-order form of the Lagrange density, the field A_0 is recognized as a Lagrange multiplier field which enforces the Chern-Simons Gauss law constraint (2.8a). This constraint may be solved by expressing the vector potential $\vec{A}$ in terms of the charge density ρ as

$$\vec{A} = -\frac{1}{\kappa}\int d^2x'\vec{G}(\vec{x},\vec{x}')\rho(\vec{x}') \qquad (2.17)$$

where the Green's function $\vec{G}(\vec{x},\vec{x}')$ is defined by[1]

$$\vec{G}(\vec{x},\vec{x}') = \vec{G}(\vec{x}-\vec{x}') \qquad (2.18a)$$

$$\vec{\nabla}\times\vec{G} \equiv \epsilon^{ij}\partial_i G_j = -\delta^2(\vec{x}) \qquad (2.18b)$$

$$G^i(\vec{x}) = -\frac{1}{2\pi}\epsilon^{ij}\partial_j \ln r \qquad (2.18c)$$

This system is therefore a constrained Hamiltonian system, with hamiltonian

$$H = \int d^2x \left(\frac{1}{2m}\left|\vec{D}\psi\right|^2 - \frac{g}{2}|\psi|^4\right) \qquad (2.19)$$

where the vector potential $\vec{A}$ which appears inside the covariant derivative $\vec{D}\psi$ is given by the expression (2.17). The hamiltonian

---

[1]Some special properties of these definitions for the Green's function have been discussed in the literature [106,130].

(2.19) is therefore a nonlocal, nonlinear expression in terms of the matter field ψ. The time evolution for the matter field ψ is obtained from the Heisenberg equation

$$i\partial_0 \psi = \frac{\delta H}{\delta \psi^*} \tag{2.20}$$

where it is important to remember that in varying H with respect to ψ^* one must also vary $\vec{A}$ with respect to ψ^* according to the relation (2.17). This generates the $A_0 \psi$ term appearing in the original matter equation of motion (2.4) with A_0 defined in terms of the matter fields as

$$A_0 = -\frac{1}{\kappa} \int d^2 x' \vec{G}(\vec{x}, \vec{x}') \cdot \vec{J}(\vec{x}') \tag{2.21}$$

Note that this relation is consistent with the 'other' gauge field equation of motion (2.8b) and with the time evolution of the constraint (2.17) defining $\vec{A}$ in terms of ρ. These equations defining the gauge fields in terms of the matter fields involve a choice of gauge (for example, the relation (2.8) determines $\vec{A}$ only up to a divergence term), an issue which we shall confront in the quantization of these systems in Chapter 6.

C. Static Self-Dual Solutions

To seek solutions to the Euler-Lagrange equations (2.4) we shall find it useful to make a "self-dual" ansatz for the matter fields [125]. Suppose the matter field ψ satisfies the self-dual ansatz:

$$D_j \psi = \pm i \epsilon_{jk} D_k \psi \tag{2.22}$$

This may be written in a more compact form

$$D_\mp \psi = 0 \tag{2.23}$$

if we introduce the characteristic coordinates

$$x^{\pm} = \frac{1}{2}(x^1 \mp ix^2) \tag{2.24a}$$

$$\partial_{\pm} = \partial_1 \pm i\partial_2 \tag{2.24b}$$

$$D_{\pm} = D_1 \pm iD_2 \tag{2.24c}$$

With the fields subject to the ansatz (2.23) (equivalently (2.22)), the current density (2.5b) simplifies to

$$J^j = \pm\frac{1}{2m}\epsilon^{jk}\partial_k\left(|\psi|^2\right) \tag{2.25}$$

and so the electric field equation of motion (2.8b) becomes:

$$\partial_i A_0 - \partial_0 A_i = \pm\frac{1}{2m\kappa}\partial_i\left(|\psi|^2\right) \tag{2.26}$$

It is also useful to record the following *factorization identity*:

$$\vec{D}^2\psi = D_{\pm}D_{\mp}\psi \mp F_{12}\psi \tag{2.27}$$

This identity, which factorizes the covariant Laplacian (producing an additional $B\psi$ term) plays a prominent role throughout these Lectures. Using the magnetic field equation of motion (2.8a) which relates F_{12} to the charge density ρ, we see that the matter field equation (2.4a), the gauged nonlinear Schrödinger equation, becomes

$$i\partial_0\psi = -\frac{1}{2m}D_{\pm}D_{\mp}\psi - \left(g \mp \frac{1}{2m\kappa}\right)|\psi|^2\psi + A_0\psi \tag{2.28}$$

With the self-dual ansatz, the first term on the RHS of (2.28) vanishes, giving

$$i\partial_0\psi = -\left(\left(g \mp \frac{1}{2m\kappa}\right)|\psi|^2 - A_0\right)\psi \tag{2.29}$$

22

By inspection, we see that the equations (2.26) and (2.29) with the self-dual ansatz (2.23) are solved by *static* solutions

$$\partial_0 \psi = 0$$

$$\partial_0 A_i = 0 \qquad (2.30)$$

with A_0 chosen as

$$A_0 = \pm \frac{1}{2m\kappa} |\psi|^2 \qquad (2.31)$$

provided the strength g of the nonlinear coupling in (2.1) is chosen to take the special critical value

$$g = \pm \frac{1}{m\kappa} \qquad (2.32)$$

which we shall refer to as the "self-dual coupling". The corresponding self- dual Lagrange density is

$$\mathcal{L} = \frac{\kappa}{2} \epsilon^{\mu\nu\rho} A_\mu \partial_\nu A_\rho + i\psi^* D_0 \psi - \frac{1}{2m} \left| \vec{D}\psi \right|^2 \pm \frac{1}{2m\kappa} |\psi|^4 \qquad (2.33)$$

We thus arrive at the "nonrelativistic self-dual Chern-Simons equations":

$$D_\mp \psi = 0 \qquad (2.34a)$$

$$F_{12} = \frac{1}{\kappa} |\psi|^2 \qquad (2.34b)$$

As described above, with the critical value (2.32) for the nonlinear coupling g, solutions to the nonrelativistic self-dual Chern- Simons equations provide *static* solutions to the Euler-Lagrange equations of motion. Also note that the self-duality equations (2.34) are *first-order* equations, rather than the second-order equations of motion (2.4). This is a familiar feature of self-dual models, as was indicated in the Introduction.

23

A clearer insight into the *physical* nature of these self-duality equations comes from considering the Hamiltonian formalism. The factorization identity (2.27) may be re-expressed as:

$$|\vec{D}\psi|^2 = |D_{\mp}\psi|^2 \pm F_{12}|\psi|^2 \mp m\epsilon^{ij}\partial_i J_j \qquad (2.35)$$

Using this identity, together with the Gauss law constraint (2.8a), we find that the Hamiltonian (2.19) is

$$H = \int d^2x \left(\frac{1}{2m}|D_{\mp}\psi|^2 - \frac{1}{2}\left(g \mp \frac{1}{m\kappa}\right)|\psi|^4 \right) \qquad (2.36)$$

where we have dropped a spatial boundary term $\frac{1}{2}\int \epsilon^{ij}\partial_i J_j$. With the critical self-dual coupling (2.32), the energy reduces to

$$H = \frac{1}{2m} \int d^2x |D_{\mp}\psi|^2 \qquad (2.37)$$

which is manifestly positive, and which is minimized by configurations satisfying the self-duality equations (2.34). Since H is minimized by the self-dual solutions, these solutions necessarily correspond to *static* solutions of the Euler-Lagrange equations of motion, as demonstrated explicitly above. In fact, the relationship between the static solutions and the self-duality equations (2.34) is deeper than is indicated here. So far we have shown that the self-dual solutions are necessarily static. In the next Section we show that the correspondence works in the other direction also: **all** static solutions are solutions of the self-duality equations.

The particular self-dual form of the potential in (2.33) may also be understood as a Pauli-like magnetic interaction. Consider a two-component spinor χ (as is appropriate for a planar theory) and let

$$S = \vec{\sigma} \cdot \left(\vec{\nabla} + i\vec{A}\right) \chi \qquad (2.38)$$

where the Pauli matrices σ^1 and σ^2 satisfy

$$\sigma^i \sigma^j = \delta^{ij} + i\epsilon^{ij} \sigma^3 \tag{2.39}$$

Then the Pauli energy is

$$E = \frac{1}{2m} \int d^2x S^\dagger S$$

$$= \frac{1}{2m} \int d^2x |\vec{D}\chi|^2 + \frac{1}{2m} \int d^2x B\chi^\dagger \sigma^3 \chi \tag{2.40}$$

Taking χ to be an eigenstate of σ^3, with eigenvalue ± 1,

$$\chi_+ = \begin{pmatrix} \psi \\ 0 \end{pmatrix} \qquad \chi_- = \begin{pmatrix} 0 \\ \psi \end{pmatrix} \tag{2.41}$$

the Pauli energy (2.40) becomes

$$E = \frac{1}{2m} \int d^2x |\vec{D}\psi|^2 \pm \frac{1}{2m} \int d^2x B|\psi|^2$$

$$= \frac{1}{2m} \int d^2x |\vec{D}\psi|^2 \pm \frac{1}{2m\kappa} \int d^2x |\psi|^4 \tag{2.42}$$

after using the Gauss law constraint (2.8a). Thus, the self-dual non-linear interaction term in (2.33) may be alternatively viewed as a magnetic moment interaction with a magnetic field that is given self-consistently in terms of the charge density ρ by the Chern- Simons equation (2.8a). This planar Pauli system is well-known to be an example of supersymmetric quantum mechanics [96,38]. The role of supersymmetry is in fact much deeper, as the self-dual model with Lagrange density (2.33) may be embedded into a theory possessing a super-Galilean invariance [161]. Also, the scalar matter-Chern-Simons self-dual models described here may be generalized to spinor theories with an analogous self-dual structure [69].

D. Dynamical Symmetries

As expected, the nonrelativistic field theory with Lagrange density (2.1) possesses the *kinematic* symmetry of Galilean invariance. However, we shall see below that there is in addition a further *dynamical* invariance under conformal reparametrizations of time [133,134].

The standard Galilean transformations are:

1. time translation:

$$t \to t' = t + a$$

$$\vec{x} \to \vec{x}' = \vec{x} \qquad (2.43)$$

2. space translation:

$$t \to t' = t$$

$$\vec{x} \to \vec{x}' = \vec{x} + \vec{a} \qquad (2.44)$$

3. space rotation:

$$t \to t' = t$$

$$x_i \to x_i' = \mathcal{R}_{ij}(\omega)x_j \qquad (2.45)$$

where $\mathcal{R}_{ij}(\omega)$ is the rotation matrix for a rotation through angle ω.

4. Galilean boost:

$$t \to t' = t$$

$$\vec{x} \to \vec{x}' = \vec{x} + \vec{v}t \qquad (2.46)$$

Under the first three types of Galilean transformation (time translation, space translation and space rotation) the matter field ψ transforms as a scalar

$$\psi'(t', \vec{x}') = \psi(t, \vec{x}) \qquad (2.47)$$

but, as is well known in nonrelativistic quantum theory, under a Galilean boost (2.46) the field ψ transforms with a 1-cocycle

$$\psi'(t', \vec{x}') = e^{im\vec{v}\cdot(\vec{x}+\vec{v}t/2)}\psi(t, \vec{x}) \qquad (2.48)$$

The nonrelativistic Chern-Simons system with Lagrange density (2.1) has additional invariances, beyond the Galilean symmetries (2.43,2.44,2.45,2.46), corresponding to conformal reparametrizations of the time coordinate. Consider the special class of coordinate tranformations

$$t \to t' = T(t)$$

$$\vec{x} \to \vec{x}' = \sqrt{\dot{T}}\,\vec{x} \qquad (2.49)$$

in which the time coordinate is reparametrized, and the spatial coordinates are multiplied by an associated time dependent factor. Define the transformed field as

$$\psi'(t', \vec{x}') = \frac{1}{\sqrt{\dot{T}}}e^{imr^2\ddot{T}/(4\dot{T})}\psi(t, \vec{x}) \qquad (2.50)$$

which includes both a Jacobian weight factor and a cocycle factor. Then if $\psi'(t', \vec{x}')$ satisfies the original gauged nonlinear Schrödinger equation (2.4a) in terms of the primed coordinates, the field $\psi(t, \vec{x})$ satisfies

$$iD_0\psi = -\frac{1}{2m}\vec{D}^2\psi - g\,|\psi|^2\,\psi + \frac{1}{2}m\omega^2(t)\vec{x}^2\psi \qquad (2.51)$$

27

This is the gauged nonlinear Schrödinger equation (2.4a) with a new term corresponding to a harmonic potential with time-dependent frequency (squared)

$$\omega^2(t) = \frac{\dddot{T}}{2\dot{T}} - \frac{3}{4}\frac{(\ddot{T})^2}{(\dot{T})^2} \tag{2.52}$$

There are three special transformations of the form (2.49) for which this frequency in (2.52) *vanishes*. These transformations therefore correspond to symmetries of the system. The first is just the time translation symmetry (2.43). The other two are

5. time dilation:

$$t \rightarrow t' = at$$

$$\vec{x} \rightarrow \vec{x}' = \sqrt{a}\,\vec{x} \tag{2.53}$$

6. special conformal time transformation:

$$\frac{1}{t} \rightarrow \frac{1}{t'} = \frac{1}{t} + a$$

$$\vec{x} \rightarrow \vec{x}' = \frac{1}{1+at}\vec{x} \tag{2.54}$$

It is straightforward to check that $\omega(t)$ given in (2.52) does indeed vanish for each of the transformations (2.43,2.53,2.54).

Under the time dilation (2.53) the field ψ transforms as

$$\psi'(t', \vec{x}') = \frac{1}{\sqrt{a}}\psi(t, \vec{x}) \tag{2.55}$$

while under the special conformal transformation (2.54), ψ transforms as

$$\psi'(t', \vec{x}') = (1+at)e^{-imar^2/2(1+at)}\psi(t, \vec{x}) \tag{2.56}$$

Note that for each of these symmetry operations, the matter density ρ transforms with a Jacobian factor

$$\rho'(t', \vec{x}') = \mathcal{J}\rho(t, \vec{x})$$

$$\mathcal{J} \equiv det\left(\frac{\partial x^i}{\partial x'^j}\right) \tag{2.57}$$

and the gauge fields A^μ, defined in (2.17) and (2.21), transform covariantly

$$A'_\mu(t', \vec{x}') = \frac{\partial x^\nu}{\partial x'^\mu}A_\nu(t, \vec{x}) \tag{2.58}$$

It is therefore straightforward to check that these transformations do indeed leave the action corresponding to the Lagrange density (2.1) invariant, and so correspond to symmetry operations. The corresponding conserved generators can be obtained from Noether's theorem.

The generators of the standard Galilean transformations (2.43,2.44,2.45,2.46) can also be found from the energy-momentum tensor. The energy density is

$$T^{00} \equiv \mathcal{E} = \frac{1}{2m}|\vec{D}\psi|^2 - \frac{g}{2}|\psi|^4 \tag{2.59}$$

The momentum density is

$$\vec{\mathcal{P}} = m\vec{J} = -\frac{i}{2}\left(\psi^\dagger \vec{D}\psi - \left(\vec{D}\psi\right)^\dagger \psi\right) \tag{2.60}$$

The energy and momentum density satisfy the continuity eqautions

$$\partial_0\mathcal{E} + \vec{\nabla} \cdot \vec{T} = 0 \tag{2.61a}$$

$$\partial_0\mathcal{P}^i + \partial_j T^{ij} = 0 \tag{2.61b}$$

29

where $\vec{T}$ is the energy flux

$$\vec{T} = -\frac{1}{2} \left((D_0 \psi)^* \, \vec{D} \psi + \left(\vec{D} \psi \right)^* D_0 \psi \right) \tag{2.62}$$

and T^{ij} is the momentum flux (or stress tensor)

$$T^{ij} = \frac{1}{2} \left((D_i \psi)^* D_j \psi + (D_j \psi)^* D_i \psi - \delta^{ij} |\vec{D}\psi|^2 \right)$$

$$+ \frac{1}{4} \left(\delta^{ij} \vec{\nabla}^2 - 2\partial_i \partial_j \right) \rho + \delta^{ij} \mathcal{E} \tag{2.63}$$

Notice that the energy flux $T^i \equiv T^{i0}$ does not equal the momentum density $\mathcal{P}^i \equiv T^{0i}$, as would be the case in a relativistic theory. However, this theory is rotationally invariant, and so the stress tensor T^{ij} is symmetric. Also note that the energy density $\mathcal{E}$ is one-half the trace of the stress tensor

$$\mathcal{E} = \frac{1}{2} \sum_{i=1}^{2} T^{ii} \tag{2.64}$$

which reflects the conformal invariance of this nonrelativistic system (contrast with a *relativistic* conformally invariant model, for which $\mathcal{E} = \sum_i T^{ii}$). This relation between the energy and the trace of the stress tensor will be important in the quantization of these systems - see Chapter 6.

The Galilean constants of motion corresponding to the symmetry transformations (2.43,2.44,2.45,2.46) are:

1. energy

$$E = \int d^2x \, \mathcal{E} \tag{2.65}$$

2. momentum

$$\vec{P} = \int d^2x \, \vec{\mathcal{P}} \tag{2.66}$$

3. angular momentum

$$M = \int d^2x \, \vec{x} \times \vec{\mathcal{P}} \qquad (2.67)$$

4. Galilean boost

$$\vec{B} = t\vec{P} - m \int d^2x \, \rho \, \vec{x} \qquad (2.68)$$

and are standard for any Galilean invariant theory. The conservation of these quantities follows directly from the continuity equations (2.61) together with the matter density continuity equation (2.11) for ρ and $\vec{J}$, which expresses the abelian phase invariance of the system.

The conserved quantities corresponding to the additional conformal transformations (2.53,2.54) are:

5. dilation

$$D = tE - \frac{1}{2} \int d^2x \, \vec{x} \cdot \vec{\mathcal{P}} \qquad (2.69)$$

6. special conformal

$$K = -t^2 E + 2tD + \frac{m}{2} \int d^2x \, \rho \, \vec{x}^2 \qquad (2.70)$$

At the classical level, using canonical Poisson brackets for the fields, one finds that the three generators E, D and K satisfy the $SO(2,1)$ commutation relations, constituting a dynamical symmetry algebra of the system. It is interesting to note that the magnetic vortex system also has an $SO(2,1)$ dynamical symmetry [124]. For a general discussion of scale invariance in physics and field theory see [121].

We stress that this dynamical conformal symmetry is present at the classical level for **any** value of the nonlinear coupling constant g appearing in the Lagrange density (2.1). Later (see Chapter 6) we shall consider the question of the fate of this classical symmetry

31

in the *quantum* theory and we shall see that this classical conformal symmetry is broken, unless g takes a critical value corresponding to the self-dual value in (2.32).

This dynamical symmetry guarantees that static solutions are necessarily self-dual. To see this, consider the dilation generator D defined in (2.69). D itself is conserved and so is time independent. For static solutions $\vec{\mathcal{P}}$ is also time independent, and so (2.69) implies that E must vanish. But from (2.37) we see that the energy vanishes only for self-dual solutions. A similar argument using the special conformal generator K in (2.70) shows that both E and D vanish for static solutions.

Further, from the Galilean boost generator $\vec{B}$ in (2.68) we see that for static solutions the momentum $\vec{P}$ vanishes (as it should!). The angular momentum for a static (self-dual) solution reduces to

$$M = \mp\frac{1}{2}\int d^2x\, x^i \partial_i \left(|\psi|^2\right) = \pm\int d^2x\, \rho = \pm Q \qquad (2.71)$$

which is proportional to the net matter density. For self-daul solutions, the Galilean boost generator in (2.68) reduces to an expression proportional to the electric dipole moment

$$\vec{B} = -m\int d^2x\, \rho\, \vec{x} \qquad (2.72)$$

while the dilation generator in (2.69) becomes

$$D = \mp\frac{1}{4}\int d^2x\, \vec{x} \times \vec{\nabla}\rho = 0 \qquad (2.73)$$

The self-dual special conformal generator in (2.70) is proportional to the electric quadrupole moment

$$K = \frac{m}{2}\int d^2x\, \rho\, \vec{x}^2 \qquad (2.74)$$

To conclude this Section on the dynamical symmetries of the non-relativistic self-dual Chern-Simons theory, we note that this system has also been formulated in a nonrelativistic Kaluza-Klein framework [67] in which the nonrelativistic $2+1$ dimensional space-time is obtained from a $3+1$ dimensional Lorentz manifold by dimensional reduction. The nonrelativistic conformal symmetries in the $2+1$ dimensional model may then be related to the relativistic symmetries of the original Lorentz manifold which survive the reduction. This interesting alternative viewpoint deserves further investigation, in particular to discover the *geometrical* significance of the *self-duality* condition.

E. Explicit Self-Dual Solutions: The Liouville Equation

The nonrelativistic self-dual Chern-Simons equations (2.34) can, in fact, be solved completely and explicitly. If the (complex) field ψ is decomposed as

$$\psi = e^{-i\omega}\rho^{1/2} \tag{2.75}$$

then the first self-duality equation (2.34a) yields the vector potential as

$$A_i = \partial_i\omega \pm \frac{1}{2}\epsilon^{ij}\partial_j ln\rho \tag{2.76}$$

away from the zeros of ρ. Inserting this form for $\vec{A}$ into the second self-duality equation (2.34b) produces the following equation for the charge density ρ

$$\vec{\nabla}^2 ln\rho = \mp\frac{2}{\kappa}\rho \tag{2.77}$$

which must be satisfied away from the zeros of ρ. We recognize (2.77) as the famous Liouville equation, which is known to be integrable and indeed completely solvable [182]. The phase ω of ψ is determined

33

by demanding $\vec{A}$ to be nonsingular in the vicinity of the zeros of ρ, as is illustrated below in detail.

To describe the general solutions to the Liouville equation (2.77) we first re-express it in terms of the characteristic coordinates (2.24)

$$\partial_+\partial_- ln\rho = \mp\frac{2}{\kappa}\rho \qquad (2.78)$$

It is a simple matter to verify that

$$\rho = \pm\kappa\partial_+\partial_- ln(1 + f(x^-)g(x^+)) \qquad (2.79)$$

satisfies the Liouville equation (2.78) for *any* functions f and g. This follows simply because

$$\rho = \pm\kappa\frac{f'(x^-)g'(x^+)}{(1 + f(x^-)g(x^+))^2} \qquad (2.80)$$

(where $f' \equiv \partial_- f$ and $g' \equiv \partial_+ g$) so that

$$ln\rho = ln(\pm\kappa) + \ln\left(f'(x^-)\right) + \ln\left(g'(x^+)\right) - 2\ln\left(1 + f(x^-)g(x^+)\right) \qquad (2.81)$$

from which (2.78) immediately follows. In fact, this solution (2.79) is the most general solution [182].

For *real* and *regular* solutions for ρ, we take $g(x^+) = (f(x^-))^*$, so that

$$\rho = \pm\kappa\nabla^2 ln\left(1 + |f(x^-)|^2\right)$$

$$= \pm\kappa\frac{|f'(x^-)|^2}{(1 + |f(x^-)|^2)^2} \qquad (2.82)$$

Since $\rho \equiv |\psi|^2$ is a charge density we require it to be positive, and so we must choose the $\pm$ sign according to the sign of κ in such a way that

34

$$\pm\kappa = |\kappa| \tag{2.83}$$

Thus, in order to obtain an appropriate, regular solution to the Liouville equation (2.77) (which in turn, yields a solution to the self-duality equations (2.34)), we see that it is necessary to correlate the choice of a self-dual or anti-self-dual ansatz in (2.23) with the sign of κ so that (2.83) is satisfied.[2] Another way to say this is that only the Liouville equation with sign

$$\vec{\nabla}^2 ln\rho = -\frac{2}{|\kappa|}\rho \tag{2.84}$$

has real positive regular solutions.

From now on, we choose the sign of κ to be *positive* in which case we obtain self-dual solutions by solving the self-duality equation (2.34a) with a D_-:

$$D_-\psi = 0 \tag{2.85}$$

It is very important to stress that with the sign correlation in (2.83), the self-dual coupling strength g in (2.33) with the critical value (2.32) is always *positive* (independent of the sign of κ), corresponding always to an *attractive* self- dual potential [125,126]

$$V_{\text{SD}} = -\frac{1}{2m|\kappa|}|\psi|^4 \tag{2.86}$$

Explicit radially symmetric solutions may be obtained by taking

---

[2]Note that if we choose the *opposite* sign correlation, $\pm\kappa = -|\kappa|$, it would still be possible to obtain positive real solutions for ρ by choosing $g(x^-) = -f(x^+)^*$, but these solutions would not be regular. We return to this possibility in Chapter 6, when we consider quantization.

$$f(x^-) = \left(\frac{x^-}{x_0^-}\right)^{-n} = \left(\frac{r}{r_0}\right)^{-n} e^{in\theta} \tag{2.87}$$

where we have introduced the polar coordinates $x^\pm = \frac{r}{2}e^{\mp i\theta}$; (note the factor of $1/2$, which enters because of the definition of the characteristic coordinates in (2.24)). The corresponding self-dual solution has charge density

$$\rho = \frac{4|\kappa|n^2}{r_0^2} \frac{(r/r_0)^{2(n-1)}}{\left(1 + (r/r_0)^{2n}\right)^2} \tag{2.88}$$

As $r \to 0$, the charge density behaves as

$$\rho \sim r^{2(n-1)} \tag{2.89}$$

while as $r \to \infty$

$$\rho \sim r^{-2-2n} \tag{2.90}$$

The vector potential for $r \to 0$ behaves as

$$A_i(r) \sim \partial_i \omega \pm (n-1)\epsilon_{ij}\frac{x^j}{r^2} \tag{2.91}$$

We can therefore avoid singularities in the the vector potential at the origin if we choose the phase of ψ to be

$$\omega = \pm(n-1)\theta \tag{2.92}$$

Thus the self-dual ψ field is

$$\psi = \frac{2n\sqrt{|\kappa|}}{r_0} \frac{(r/r_0)^{n-1}}{1 + (r/r_0)^{2n}} e^{\pm i(n-1)\theta} \tag{2.93}$$

Requiring that ψ be single-valued we find that n must be an integer, and for ρ to decay at infinity we require that n be positive. For $n > 1$ the ψ solution has vorticity n at the origin and ρ goes to zero at the origin.

36

The net matter charge Q corresponding to the solution (2.82) is

$$Q = |\kappa| \int d^2x \, \nabla^2 ln \left(1 + |f|^2\right)$$

$$= 2\pi|\kappa| \left[r\frac{d}{dr}ln \left(1 + |f|^2\right)\right]_0^\infty \tag{2.94}$$

For the radial solution (2.88), the net matter charge is

$$Q = \int d^2x \, \rho = 4\pi|\kappa|n \tag{2.95}$$

and the corresponding flux is

$$\Phi = 4\pi n \tag{2.96}$$

which represents an *even* number of flux units. This quantized character of the flux is related to a special inversion symmetry of the Liouville equation, and is not particular to radially symmetric solutions [151]. The angular momentum generator (2.67) is

$$M = -4\pi|\kappa|n \tag{2.97}$$

and the conformal charge (2.70) is

$$K = 2m\pi|\kappa|nr_0^2 \left(\frac{\pi/n}{sin\pi/n}\right) \tag{2.98}$$

The radially symmetric solution (2.93) arose from choosing the holomorphic function $f(x^-)$ in (2.82) as

$$f(x^-) \sim \frac{1}{(x^-)^n} \tag{2.99}$$

and corresponds to n solitons superimposed at the origin. A solution corresponding to n *separated* solitons may be obtained by taking

$$f(x^-) = \sum_{a=1}^n \frac{c_a}{x^- - x_a^-} \tag{2.100}$$

37

Note that there are $4n$ real parameters involved in this solution : $2n$ real parameters x_a^- $(a = 1, \ldots n)$ describing the locations of the solitons, and $2n$ real parameters c_a $(a = 1, \ldots n)$ corresponding to the scale and phase of each soliton. The solution in (2.100) has in fact been shown, by an index theory counting argument, to be the most general multi-soliton solution [150]. Solutions with a *periodic* matter density ρ may be obtained by choosing the function f in (2.82) to be a doubly periodic function [198]. Other properties of these self-dual Chern-Simons vortices, in relation to vortices in planar Maxwell electrodynamics, have been discussed in [53].

F. Time-Dependent Solutions

As discussed in Section II D, solutions to the self-duality equations (2.34) provide *all* the solutions to the *static* Euler-Lagrange equations of motion (2.4). We can then make use of the Galilean and conformal symmetries of the theory to construct explicitly time-dependent solutions based on these static solutions [131]. The most obvious of these is the Galilean boost of a static solution $\psi_{\text{static}}(\vec{x})$:

$$\psi(t, \vec{x}) = e^{-im\vec{v} \cdot (\vec{x} + \vec{v}t/2)} \psi_{\text{static}}(\vec{x} + \vec{v}t) \qquad (2.101)$$

The corresponding density

$$\rho(t, \vec{x}) = \rho_{\text{static}}(\vec{x} + \vec{v}t) \qquad (2.102)$$

corresponds to a translation, with uniform velocity $\vec{v}$, of the static density $\rho_{\text{static}}(\vec{x})$. As such, this time-dependent solution is not particularly interesting.

More interesting is the time dependent solution obtained by applying the conformal transformation (2.54) to a static solution:

38

$$\psi(t, \vec{x}) = \frac{1}{1 + at} e^{imar^2/2(1+at)} \psi_{\text{static}} \left(\frac{\vec{x}}{1 + at} \right) \qquad (2.103)$$

This leads to a time-dependent density

$$\rho(t, \vec{x}) = \frac{1}{(1 + at)^2} \rho_{\text{static}} \left(\frac{\vec{x}}{1 + at} \right) \qquad (2.104)$$

The conserved generators (2.65,2.66,2.67, 2.68,2.69,2.70) may be evaluated for these time dependent solutions. For the Galilean boosted solution (2.101)

$$E = \frac{1}{2} m \vec{v}^2 Q$$

$$\vec{P} = m \vec{v} Q$$

$$M = Q \left(1 + m \vec{v} \times < \vec{x} > \right)$$

$$D = -\frac{1}{2} \int d^2 x \vec{x} \cdot \vec{P} - \frac{1}{2} m \vec{v} \cdot < \vec{x} > \qquad (2.105)$$

where $< \vec{x} >$ denotes the expectation value

$$< \vec{x} > = \frac{\int d^2 x \rho \vec{x}}{\int d^2 x \rho} \qquad (2.106)$$

Both the Galilean boost generator $\vec{B}$ and the conformal generator K are unchanged for this time dependent solution. The generators in (2.105) illustrate the fact that the boosted soliton behaves like a particle of total mass mQ.

For the conformally boosted solution (2.103) one finds

$$E = \frac{1}{2} m a^2 Q < \vec{x}^2 >$$

$$\vec{P} = m a Q < \vec{x} >$$

39

$$D = -\frac{1}{2}maQ < \vec{x}^2 > \qquad (2.107)$$

with $\vec{B}$ and K unchanged. Finally, one may combine the Galilean boost and the special conformal transformation (the order is irrelevant since $\vec{B}$ and K commute as generators) to obtain even more general time dependent solutions [131]. The existence properties of these formal time dependent solutions have recently been investigated in the context of the initial value problem for the nonrelativistic self-dual Chern-Simons system [19]. Based on the relation (2.70) (which is valid for all time)

$$\frac{m}{2} \int d^2x \, \rho \, \vec{x}^2 = Et^2 - 2Dt + K \qquad (2.108)$$

and the manifest positivity of the LHS for all time, one finds further conditions on E, D and K which must be satisfied for consistent time evolution. In particular, for certain initial data, the time dependent solutions are found to collapse in a finite time [19].

G. Solutions in External Fields

The time-dependent solutions described in Section II F were obtained by transforming the static solutions according to a **symmetry** operation of the system, which leaves the action invariant. Alternatively, one may make a transformation of the action which **changes** the action to the action for a new model. Then solutions of the original model may be transformed to solutions (in general, now time dependent solutions) of the new model. This is, of course, only a useful exercise if the transformed system is an interesting one. Fortunately, it is a property of nonrelativistic dynamics that a nonrelativistic action may be easily transformed so as to introduce an additional harmonic potential $V \sim \frac{1}{2}\omega^2 \vec{x}^2$. In two spatial

40

dimensions this is of direct interest for two reasons. First, the harmonic potential may be used as a confining potential which permits an elegant and simple treatment of the thermodynamic and statistical mechanical properties of many-particle systems [191,37,43]. Second, a minor variant of this transformation amounts to the introduction of an external uniform magnetic field transverse to the spatial plane, a configuration which is of great phenomenological interest for applications of Chern-Simons theories to the quantum Hall effect [78–81,175].

The idea is most easily introduced with the harmonic potential case, and without coupling to gauge fields. Suppose $\psi(\vec{x})$ is a static solution of the nonlinear[3] Schrödinger equation with zero energy

$$\left(-\frac{1}{2m}\vec{\nabla}^2 - g|\psi|^2\right)\psi(\vec{x}) = 0 \qquad (2.109)$$

Then the time dependent function

$$\psi^{(\omega)}(t,\vec{x}) = \frac{1}{\cos\omega t}e^{-i(m/2)\omega\vec{x}^2\tan\omega t}\psi\left(\frac{\vec{x}}{\cos\omega t}\right) \qquad (2.110)$$

satisfies the Schrödinger equation

$$i\partial_t\psi^{(\omega)} = \left(-\frac{1}{2m}\vec{\nabla}^2 - g|\psi^{(\omega)}|^2 + \frac{m}{2}\omega^2\vec{x}^2\right)\psi^{(\omega)} \qquad (2.111)$$

This corresponds to choosing the transformation function

$$T(t) = \frac{1}{\omega}\tan\omega t \qquad (2.112)$$

in the coordinate transformation (2.49). Then the transformed wavefunction acquires a cocycle and a Jacobian weight factor as in (2.50), and the frequency (2.52) of the new harmonic term generated in (2.51) is constant

---

[3]This argument works both with and without the nonlinear term.

41

$$\frac{\ddot{T}}{2\dot{T}} - \frac{3}{4}\frac{(\ddot{T})^2}{(\dot{T})^2} = \omega^2 \tag{2.113}$$

Note that the solution (2.110) is periodic in time. This periodicity is important for the semi-classical quantization of these classical solutions [131,132].

This transformation generalizes to the gauged nonlinear Schrödinger equation (2.4a). That is, suppose $\psi(\vec{x})$ satisfies

$$\left(-\frac{1}{2m}\left(\vec{\nabla} + i\vec{A}\right)^2 - g|\psi|^2\right)\psi(\vec{x}) = 0 \tag{2.114}$$

where the vector potential $\vec{A}$ is given by the expression (2.17). Then the transformed field $\psi(t,\vec{x})$ in (2.110) satisfies

$$i\left(\partial_t + iA_0^{(\omega)}\right)\psi^{(\omega)} = \left(-\frac{1}{2m}\left(\vec{\nabla} + i\vec{A}^{(\omega)}\right)^2 - g|\psi^{(\omega)}|^2 + \frac{m}{2}\omega^2\vec{x}^2\right)\psi^{(\omega)}$$

$$\tag{2.115}$$

where

$$A_0^{(\omega)} = -\frac{1}{\kappa}\int d^2x\,\vec{G}\cdot\vec{J}^{(\omega)}$$

$$\vec{A}^{(\omega)} = -\frac{1}{\kappa}\int d^2x\,\vec{G}\rho^{(\omega)} \tag{2.116}$$

Evaluating the conserved energy and angular momentum generators on the time-dependent solutions we find

$$E^{(\omega)} = \omega^2 K$$

$$M^{(\omega)} = M \tag{2.117}$$

where K and M refer to the corresponding generators for the solutions without the harmonic potential.

This idea can also be extended to the case of an external uniform magnetic field, of magnitude $\mathcal{B}$ perpendicular to the plane, by defining

$$\psi^{(\mathcal{B})}(t,\vec{x}) = \frac{e^{-i\mathcal{B}r^2 tan(\mathcal{B}t/2m)/4m}e^{-iQ\mathcal{B}t/(4\pi\kappa m)}}{cos(\mathcal{B}t/2m)}\psi\left(\frac{\mathcal{R}^{ij}\left(\mathcal{B}t/2m\right)x^j}{cos(\mathcal{B}t/2m)}\right)$$

$$(2.118)$$

which corresponds to a time-dependent dilation, a time-dependent rotation through angle $\mathcal{B}t/2m$, and a gauge transformation. Here Q is the net charge of the untransformed solution. Then if $\psi(\vec{x})$ is a solution of the original gauged nonlinear Schrödinger equation (2.4a), then $\psi^{(\mathcal{B})}(t,\vec{x})$ satisfies the gauged nonlinear Schrödinger equation with an additional external magnetic field:

$$i\left(\partial_t + iA_0^{(\mathcal{B})}\right)\psi^{(\mathcal{B})} = \left(-\frac{1}{2m}\left(\vec{\nabla} + i\vec{A}^{(\mathcal{B})}\right)^2 - g|\psi^{(\mathcal{B})}|^2\right)\psi^{(\mathcal{B})} \quad (2.119)$$

where

$$A_0^{(\mathcal{B})} = -\frac{1}{\kappa}\int d^2x\vec{G}\cdot\vec{J}^{(\mathcal{B})}$$

$$A_i^{(\mathcal{B})} = -\frac{1}{\kappa}\int d^2x G_i\rho^{(\mathcal{B})} + \frac{\mathcal{B}}{2}\epsilon_{ij}x^j \quad (2.120)$$

The corresponding conserved quantities for this new system (which includes the magnetic field $\mathcal{B}$) may be expressed in terms of the conserved generators of the original system without $\mathcal{B}$ as follows:

$$E^{(\mathcal{B})} = \frac{\mathcal{B}^2}{4m^2}K - \frac{\mathcal{B}}{2m}M$$

$$\mathcal{P}_i^{(\mathcal{B})} = \mathcal{P}_i + \frac{\mathcal{B}}{2m}\epsilon_{ij}B^j$$

$$M^{(\mathcal{B})} = M \quad (2.121)$$

As a further generalization of these solutions, rather than transforming the static (self-dual) solutions (2.93) of the original model without the harmonic potential (or magnetic field), one may transform the boosted time dependent solutions (2.101) or (2.103) described

43

in section II F. These time dependent solutions may be used in a semiclassical quantization of these self-dual Chern-Simons systems [131,132]. Furthermore, the external field solutions discussed in this Section can also be found in the nonrelativistic Kaluza-Klein approach of Duval et al [68].

H. Maxwell–Chern–Simons Model

The nonrelativistic self-dual Chern-Simons system described in this Chapter may be generalized to include a Maxwell term for the gauge field in the Lagrange density [57]. This addition has a dramatic effect on the model because the gauge field now acquires a physical propagating massive mode - whereas without the Maxwell term the gauge field is itself nondynamical, being determined by the matter fields as in (2.17) and (2.21). This also means that the gauge equations of motion become *second order* equations, in contrast to the *first order* equations (2.4b) for the pure Chern-Simons case. Being first order already, the pure Chern-Simons equations of motion did not need to be factorized in order to produce first-order self-duality equations. Thus the self-duality equations (2.34) consist of the factored matter equation and the (already first order) Gauss law constraint. With a Maxwell term included for the gauge field the gauge equations must also be factored, in addition to the (new) factorization of the matter equation. It is a remarkable sign of robustness that the Chern-Simons system still admits a self-dual formulation, provided the addition of the Maxwell gauge term is accompanied by the inclusion of an extra real scalar field of mass equal to the mass of the propagating gauge mode. This is demonstrated below.

Consider the Lagrange density

$$\mathcal{L} = -\frac{1}{4e^2} F_{\mu\nu} F^{\mu\nu} + \frac{\mu}{2e^2} \epsilon^{\mu\nu\rho} A_\mu \partial_\nu A_\rho + i\psi^* D_0 \psi - \frac{1}{2m} \left| \vec{D}\psi \right|^2$$

$$-\frac{1}{2e^2} \partial_\mu N \partial^\mu N - \frac{\mu^2}{2e^2} N^2 - \frac{e^2}{8m^2} |\psi|^4 + \left(1 + \frac{\mu}{2m} \right) |\psi|^2 N \quad (2.122)$$

Note that e^2 has dimensions of mass in $2+1$ dimensions, and the dimensionless Chern-Simons coupling parameter has been written as

$$\kappa = \frac{\mu}{e^2} \quad (2.123)$$

where μ is the mass of the propagating gauge mode. The mass of the neutral scalar field N has been chosen to be equal to this gauge mass

$$m_N = \mu = m_{\text{gauge}} \quad (2.124)$$

The potential terms in (2.122) have been chosen to satisfy the self-duality requirement, as we show below. Notice that these self-dual potential terms involve interactions between the nonrelativistic scalar density $\rho = |\psi|^2$ and the neutral scalar field N.

The pure Chern-Simons limit is achieved by the combined limit in which both mass scales e^2 and μ become infinite, but in such a way that their ratio κ remains fixed:

$$e^2 \to \infty$$

$$\mu \to \infty$$

$$\kappa = \frac{\mu}{e^2} = \text{fixed} \quad (2.125)$$

In this limit, the neutral scalar field is forced to

$$N = \frac{1}{2m\kappa} \rho \quad (2.126)$$

in which case the potential becomes

45

$$-\frac{\rho^2}{2m\kappa} \qquad (2.127)$$

which is the self-dual quartic potential in (2.33) for the pure Chern-Simons self-dual model. The Lagrange density (2.122) may also be obtained as a nonrelativistic limit [57] of a relativistic self-dual Maxwell-Chern-Simons theory [166] and the massive neutral scalar field N is essential for such a limit to be well-defined. This relativistic Maxwell-Chern-Simons theory is described in Section IV G.

The Euler-Lagrange equations of motion for the system (2.122) are

$$iD_0\psi = -\frac{1}{2m}\vec{D}^2\psi - \left(1 + \frac{\mu}{2m}\right)N\psi + \frac{\mu}{4m^2\kappa}|\psi|^2\psi \qquad (2.128a)$$

$$\left(-\partial_\mu\partial^\mu + \mu^2\right)N = \frac{\mu}{\kappa}\left(1 + \frac{\mu}{2m}\right)|\psi|^2 \qquad (2.128b)$$

$$\frac{1}{e^2}\partial_\mu F^{\mu\nu} + \frac{\mu}{2e^2}\epsilon^{\nu\alpha\beta}F_{\alpha\beta} = J^\nu \qquad (2.128c)$$

where J^ν is the nonrelativistic matter current defined in (2.5). In particular, the Gauss law constraint now reads

$$\vec{\nabla}\cdot\vec{E} = \mu B - e^2\rho \qquad (2.129)$$

where B and $\vec{E}$ are the magnetic and electric fields defined in (2.9) and (2.10).

The energy density is

$$\mathcal{E} = \frac{1}{2e^2}\left(\vec{E}^2 + B^2\right) + \frac{1}{2m}\left|\vec{D}\psi\right|^2 + \frac{1}{2e^2}\left(\partial_0 N\right)^2 + \frac{1}{2e^2}\left(\vec{\nabla}N\right)^2$$

$$+\frac{\mu^2}{2e^2}N^2 - \left(1 + \frac{\mu}{2m}\right)N|\psi|^2 + \frac{e^2}{8m^2}|\psi|^4 \qquad (2.130)$$

For static fields we take $\partial_0 N = 0$ and the Gauss law (2.129) implies that we can identify (up to surface terms)

46

$$\frac{1}{2e^2}\vec{E}^2 = -\frac{\mu}{2\kappa}\left(\kappa B - \rho\right)\frac{1}{\nabla^2}\left(\kappa B - \rho\right) \qquad (2.131)$$

Furthermore, using the factorization identity (2.35) we can express the energy density as

$$\mathcal{E} = \frac{1}{2m}|D_-\psi|^2 + \frac{1}{2e^2}B\left(1 - \frac{\mu^2}{\nabla^2}\right)B + \frac{1}{2e^2}N\left(\mu^2 - \nabla^2\right)N$$

$$-\frac{e^2}{2}\rho\frac{1}{\nabla^2}\rho + \mu B\frac{1}{\nabla^2}\rho + \frac{1}{2m}B\rho$$

$$-\left(1 + \frac{\mu}{2m}\right)N\rho + \frac{e^2}{8m^2}\rho^2 \qquad (2.132)$$

which may be factorized as

$$\mathcal{E} = \left[N - e^2\left(\frac{1 + \frac{\mu}{2m}}{\mu^2 - \nabla^2}\right)\rho\right]\left(\frac{\mu^2 - \nabla^2}{2e^2}\right)\left[N - e^2\left(\frac{1 + \frac{\mu}{2m}}{\mu^2 - \nabla^2}\right)\rho\right]$$

$$+\left[B - \frac{e^2}{2m}\left(\frac{2m\mu + \nabla^2}{\mu^2 - \nabla^2}\right)\rho\right]\left(\frac{\mu^2 - \nabla^2}{-2e^2\nabla^2}\right)\left[B - \frac{e^2}{2m}\left(\frac{2m\mu + \nabla^2}{\mu^2 - \nabla^2}\right)\rho\right]$$

$$+\frac{1}{2m}|D_-\psi|^2 \qquad (2.133)$$

The energy density is bounded below by zero, and this bound is saturated by fields satisfying the following equations

$$D_-\psi = 0 \qquad (2.134a)$$

$$\left(\mu^2 - \nabla^2\right)B = \frac{e^2}{2m}\left(2m\mu + \nabla^2\right)\rho \qquad (2.134b)$$

$$\left(\mu^2 - \nabla^2\right)N = e^2\left(1 + \frac{\mu}{2m}\right)\rho \qquad (2.134c)$$

Note that the third equation (2.134c) is just the Euler-Lagrange equation of motion (2.128b) which determines the auxiliary neutral scalar field N in terms of the scalar density ρ. The first equation (2.134a) implies that

$$B = -\frac{1}{2}\nabla^2 ln\rho \qquad (2.135)$$

so that the self-duality equations (2.134) reduce to the following nonlocal equation for ρ:

$$\nabla^2 ln\rho = -\frac{e^2}{m}\left(\frac{2m\mu + \nabla^2}{\mu^2 - \nabla^2}\right)\rho \qquad (2.136)$$

In the pure Chern-Simons limit (2.125), this expression reduces to

$$\nabla^2 ln\rho = -\frac{2}{\kappa}\rho \qquad (2.137)$$

which we recognize as the Liouville equation (2.77) for the pure Chern-Simons solitons. While the general Maxwell-Chern-Simons equation (2.136) cannot be solved in closed form, it is possible to describe the global and asymptotic properties of solutions. This is due to the fact that the Chern-Simons dynamics dominates the long-distance physics. Solutions with and without vorticity are described in [57]. Some rigorous mathematical results for this system are presented in [227], where it is shown that there exist doubly periodic condensate solutions for which the number of vortices in a periodic lattice cell can be arbitrary. These are analogous to the condensate solutions studied in [198] for the pure Chern-Simons system.

III. NONABELIAN NONRELATIVISTIC MODEL

This Chapter deals with the nonabelian generalization of the abelian nonrelativistic self-dual Chern-Simons models discussed in Chapter 2. This nonabelian generalization is of interest for the understanding of nonabelian fractional statistics, a rich subject with potential applications in various quantum Hall systems [257,31]. As in the abelian models, there is a Bogomol'nyi lower bound for the energy which is saturated by static zero energy configurations which satisfy a set of first order self-duality equations. These self-duality equations may be reduced, by certain algebraic ansatze, to many examples of two dimensional integrable partial differential equations such as the classical Toda or affine Toda equations. Furthermore, with adjoint matter coupling the self-duality equations may be converted into the (Euclidean) two dimensional chiral model equation by a judicious choice of gauge. This fact leads to a complete classification of finite charge solutions in terms of Uhlenbeck's classification of finite action solutions to the chiral model (or "harmonic map") equations. This analysis is very similar in both spirit and style to the classification of finite action ("instanton") solutions to the self-dual Yang-Mills equations (1.3) in four dimensional Euclidean space. In fact, the nonabelian nonrelativistic self-dual Chern-Simons equations may be obtained from the four dimensional self-dual Yang-Mills equations by a dimensional reduction. These results lead to an interesting new relationship between the Toda and chiral model systems.

A. Nonrelativistic Self-Dual Chern–Simons Equations:

General Matter Coupling

The nonrelativistic self-dual Chern-Simons system discussed in Chapter 2 may be generalized from an abelian theory to a non-abelian theory [103,58]. We consider now a *multiplet* Ψ of complex scalar fields which transform under nonabelian local gauge transformations according to some definite representation $\mathcal{R}$ of the gauge algebra $\mathcal{G}$. (In general, $\mathcal{G}$ could be any compact simple Lie algebra, but for ease of presentation we shall primarily focus on the $SU(N)$ case. Noncompact groups, in particular $SL(2,\mathbf{R})$ and $ISO(2,1)$, have been considered in [30].) The abelian gauge field A_μ generalizes to the fields A_μ^a, with $a = 1 \ldots D$, where D is the dimension of the gauge algebra. The Lagrange density (2.33) becomes

$$\mathcal{L} = \frac{\kappa}{2}\mathcal{L}_{\text{CS}} + i\Psi^\dagger D_0 \Psi - \frac{1}{2m}(D_i\Psi)^\dagger D_i\Psi + \frac{1}{2m\kappa}\rho^a\rho^a \qquad (3.1)$$

where

$$\rho^a = -i(\Psi^\dagger \mathcal{T}^a \Psi) \qquad (3.2)$$

with $\mathcal{T}^a$ being antihermitean generators of the gauge algebra, in the representation $\mathcal{R}$ corresponding to the matter fields Ψ. The matter and gauge fields are minimally coupled through the covariant derivative

$$D_\mu\Psi = \partial_\mu\Psi + A_\mu^a \mathcal{T}^a\Psi \qquad (3.3)$$

The nonabelian Chern-Simons Lagrange density $\mathcal{L}_{\text{CS}}$ is

$$\mathcal{L}_{\text{CS}} = \epsilon^{\mu\nu\rho}\left(A_\mu^a\partial_\nu A_\rho^a + \frac{1}{3}f^{abc}A_\mu^a A_\nu^b A_\rho^c\right) \qquad (3.4)$$

where f^{abc} denotes the structure constants of the gauge algebra:

$$[\mathcal{T}^a, \mathcal{T}^b] = f^{abc}\mathcal{T}^c \tag{3.5}$$

The coefficient of the nonlinear $\rho^a \rho^a$ term in the Lagrange density (3.1) has been chosen to take the self-dual value 2.32. Without loss of generality, we have chosen the Chern-Simons coupling parameter κ to be *positive*, yielding an attractive self-dual potential[4]

$$V_{\text{SD}} = -\frac{1}{2m\kappa}\rho^a \rho^a \tag{3.6}$$

While the Chern-Simons Lagrange density (3.4) is not gauge invariant (under a gauge transformation it changes by a total derivative term and a topological term - see Equation (1.22)), its variation with respect to the gauge fields A^a_μ produces gauge covariant equations of motion. The resulting Euler-Lagrange equations are

$$iD_0\Psi = -\frac{1}{2m}\vec{D}^2\Psi + \frac{1}{m\kappa}(\Psi^\dagger \mathcal{T}^a \Psi)\mathcal{T}^a\Psi \tag{3.7a}$$

$$F^a_{\mu\nu} = -\frac{1}{\kappa}\epsilon_{\mu\nu\rho}J^{a\rho} \tag{3.7b}$$

where

$$F^a_{\mu\nu} = \partial_\mu A^a_\nu - \partial_\nu A^a_\mu + f^{abc} A^b_\mu A^c_\nu \tag{3.8}$$

is the nonabelian gauge curvature and $J^{a\mu} \equiv (\rho^a, \vec{J}^a)$ is a Lorentz covariant shorthand for the nonrelativistic matter current with ρ^a as in (3.2) and

$$\vec{J}^a = -\frac{1}{2m}\left(\Psi^\dagger \mathcal{T}^a \vec{D}\Psi - \left(\vec{D}\Psi\right)^\dagger \mathcal{T}^a \Psi\right) \tag{3.9}$$

----

[4]If κ were negative, then the critical coupling in (2.32) would still yield an attractive potential, but the notion of self-duality and anti-self-duality would be interchanged, just as in the abelian case - see Section II E.

Note that $J^{a\mu}$ satisfies the gauge covariant 'continuity equation':

$$\partial_0 \rho^a + f^{abc} A_0^b \rho^c + \vec{\nabla} \cdot \vec{J}^a + f^{abc} \vec{A}^b \cdot \vec{J}^c = 0 \qquad (3.10)$$

In addition to the gauge current $J^{a\mu}$ there is an abelian current Q^μ, corresponding to the *global* $U(1)$ symmetry of the Lagrange density (3.1):

$$Q = \Psi^\dagger \Psi \qquad (3.11a)$$

$$\vec{Q} = -\frac{i}{2m} \left(\Psi^\dagger \vec{D} \Psi - \left(\vec{D} \Psi \right)^\dagger \Psi \right) \qquad (3.11b)$$

which satisfies the ordinary continuity equation

$$\dot{Q} + \vec{\nabla} \cdot \vec{Q} = 0 \qquad (3.12)$$

The Chern-Simons equation of motion (3.7b) may be expressed in terms of the usual nonabelian magnetic and electric fields B^a and $\vec{E}^a$ as:

$$B^a \equiv F_{12}^a = \frac{1}{\kappa} \rho^a \qquad (3.13a)$$

$$E_i^a \equiv F_{i0}^a = -\frac{1}{\kappa} \epsilon_{ij} J^{aj} \qquad (3.13b)$$

To seek solutions to the Euler-Lagrange equations (3.7) we make a self-dual ansatz for the fields, as in (2.22,2.23),

$$D_- \Psi = 0 \qquad (3.14)$$

The nonabelian version of the identity (2.27) is

$$\vec{D}^2 \Psi = D_+ D_- \Psi + i F_{12}^a \mathcal{T}^a \Psi \qquad (3.15)$$

Therefore, the nonabelian gauged nonlinear Schrödinger equation (3.7a) becomes

$$iD_0\Psi = \frac{1}{2m\kappa}(\Psi^\dagger\mathcal{T}^a\Psi)\mathcal{T}^a\Psi \qquad (3.16)$$

where we have used (3.15), together with the self-dual ansatz (3.14) and the Gauss law constraint (3.13a). This can be solved by *static* solutions

$$\partial_0\Psi = 0 \qquad (3.17)$$

with A_0 given by

$$A_0^a = -\frac{i}{2m\kappa}\Psi^\dagger\mathcal{T}^a\Psi = \frac{\rho^a}{2m\kappa} \qquad (3.18)$$

Note that these solutions are consistent with the electric field equation (3.13b) because for self-dual fields satisfying (3.14) the nonrelativistic current density (3.9) simplifies to

$$J^{ai} = -\frac{i}{2m}\epsilon^{ij}\left(\partial_j\left(\Psi^\dagger\mathcal{T}^a\Psi\right) + f^{abc}A_j^b\left(\Psi^\dagger\mathcal{T}^c\Psi\right)\right) \qquad (3.19)$$

in which case the electric field equation of motion (3.13b) becomes

$$\partial_i A_0^a + f^{abc}A_i^b A_0^c = -\frac{i}{2m\kappa}\left(\partial_i\left(\Psi^\dagger\mathcal{T}^a\Psi\right) + f^{abc}A_i^b\left(\Psi^\dagger\mathcal{T}^c\Psi\right)\right) \qquad (3.20)$$

Therefore, static solutions of the Euler-Lagrange equations of motion (3.7) correspond to solutions of the first-order self-duality equations:

$$D_-\Psi = 0 \qquad (3.21a)$$

$$F_{+-}^a = -\frac{2}{\kappa}\Psi^\dagger\mathcal{T}^a\Psi \qquad (3.21b)$$

which are the natural nonabelian generalization of the abelian nonrelativistic self-dual Chern-Simons equations (2.34).

As in the abelian case, these self-duality equations may also be understood as equations for the minimization of the energy functional. The energy corresponding to the Lagrange density (3.1) is

53

$$E = \frac{1}{2m} \int d^2x (D_i\Psi)^\dagger (D_i\Psi) - \frac{1}{2m\kappa} \int d^2x \rho^a \rho^a \qquad (3.22)$$

supplemented by the Gauss law constraint (3.13a). Using the identity

$$(D_i\Psi)^\dagger(D_i\Psi) = (D_-\Psi)^\dagger(D_-\Psi) - iF_{12}^a \left(\Psi^\dagger T^a \Psi\right) - m\epsilon^{ij}\partial_i Q_j \quad (3.23)$$

we may write the energy (dropping unimportant surface terms) as

$$E = \frac{1}{2m} \int d^2x (D_-\Psi)^\dagger D_-\Psi \qquad (3.24)$$

We stress that this factorization of the energy density is only possible with the nonlinear potential term in (3.6) chosen to have the special self-dual coupling coefficient $-1/(2m\kappa)$. With this self-dual coupling, solutions to the self-duality equations (3.21) minimize the energy of the model (in fact, they correspond to *zero* energy), and so do indeed correspond to *static* solutions to the equations of motion.

This nonabelian system also possesses a dynamical $SO(2,1)$ symmetry in a manner precisely analogous to the dynamical $SO(2,1)$ symmetry of the abelian system, as described in Section II D. In fact, as in the abelian case, the existence of this dynamical symmetry at the classical level does not depend on the fact that we have chosen the strength of the nonlinear coupling in the Lagrange density (3.1) to take its self-dual value (2.32). A consequence of this dynamical symmetry is that all static solutions necessarily have zero energy, and hence are necessarily self-dual [58]. Thus, the self-dual solutions exhaust all static solutions and *vice versa*.

B. Nonrelativistic Self-Dual Chern–Simons Equations: Adjoint Matter Coupling

If the matter fields Ψ are chosen to transform according to the adjoint representation (which has dimension equal to the dimension

54

of the gauge algebra), then the nonabelian model described in Section III A takes an especially elegant form [58,59]. In the adjoint representation, the Lie algebra generators are

$$(\mathcal{T}^a)_{bc} = f^{abc} \tag{3.25}$$

and we can define a Lie algebra-valued matter field

$$\psi \equiv \Psi^a T^a \tag{3.26}$$

where the T^a are generators of the gauge Lie algebra in *any* representation. The gauge fields A_μ^a may also be expressed in terms of this *same* representation

$$A_\mu = A_\mu^a T^a \tag{3.27}$$

With this notation, the gauge covariant derivative (3.3) becomes

$$D_\mu \psi = \partial_\mu \psi + [A_\mu, \psi] \tag{3.28}$$

with the fields ψ and A_μ now being Lie algebra valued fields, taking values in the same (arbitrary) representation of the gauge Lie algebra. If the (antihermitean) generators T^a are normalized with traces

$$\mathrm{tr}\left(T^a T^b\right) = -\delta^{ab} \tag{3.29}$$

then the self-dual Lagrange density (3.1) takes the form

$$\mathcal{L} = -\frac{\kappa}{2}\mathcal{L}_{CS} + i\,tr\left(\psi^\dagger D_0 \psi\right) - \frac{1}{2m}tr\left((D_i \psi)^\dagger D_i \psi\right)$$

$$+ \frac{1}{2m\kappa}tr\left([\psi, \psi^\dagger]^2\right) \tag{3.30}$$

where

$$\mathcal{L}_{CS} = \epsilon^{\mu\nu\rho}\mathrm{tr}\left(A_\mu \partial_\nu A_\rho + \frac{2}{3}A_\mu A_\nu A_\rho\right) \tag{3.31}$$

is the Chern-Simons Lagrange density (3.4).

The Euler-Lagrange equations of motion that follow from the nonrelativistic self-dual Chern-Simons Lagrange density (3.30) are:

$$iD_0\psi = -\frac{1}{2m}\vec{D}^2\psi - \frac{1}{m\kappa}[\,[\psi,\psi^\dagger],\psi]$$

$$(3.32a)$$

$$F_{\mu\nu} = -\frac{i}{\kappa}\epsilon_{\mu\nu\rho}J^\rho$$

$$(3.32b)$$

where

$$F_{\mu\nu} = \partial_\mu A_\nu - \partial_\nu A_\mu + [A_\mu, A_\nu]$$

$$(3.33)$$

is the gauge curvature, and J^ρ is the hermitean covariantly conserved $(D_\mu J^\mu = 0)$ nonrelativistic matter current

$$J^0 = [\psi, \psi^\dagger]$$

$$(3.34a)$$

$$J^i = -\frac{i}{2m}\left([\psi^\dagger, D_i\psi] - [(D_i\psi)^\dagger, \psi]\right)$$

$$(3.34b)$$

The corresponding abelian current Q^ρ

$$Q^0 = tr\left(\psi\psi^\dagger\right)$$

$$(3.35a)$$

$$Q^i = -\frac{i}{2m}tr\left(\psi^\dagger D_i\psi - (D_i\psi)^\dagger\psi\right)$$

$$(3.35b)$$

which is ordinarily conserved $(\partial_\mu Q^\mu = 0)$. The matter field equation of motion (3.32a) is referred to as the *gauged planar nonlinear Schrödinger equation*. Note that the equations of motion (3.32) are gauge covariant with the fields transforming as

$$A_\mu \to g^{-1}A_\mu g + g^{-1}\partial_\mu g$$

$$\psi \rightarrow g^{-1}\psi g \qquad (3.36)$$

The energy density corresponding to the Lagrange density (3.30) is

$$\mathcal{E} = \frac{1}{2m}tr\left((D_i\psi)^\dagger D_i\psi\right) - \frac{1}{2m\kappa}tr\left([\psi,\psi^\dagger]^2\right) \qquad (3.37)$$

supplemented by the Gauss law constraint

$$[\psi,\psi^\dagger] = -i\kappa F_{12} \qquad (3.38)$$

which is the zeroth component of the gauge equations of motion (3.32b). To obtain the Bogomol'nyi - style lower bound for the energy density we employ the identity (3.23) with adjoint coupling,

$$tr\left((D_i\psi)^\dagger D_i\psi\right) = tr\left((D_-\psi)^\dagger D_-\psi\right) - i\,tr\left(\psi^\dagger[F_{12},\psi]\right) - m\epsilon^{ij}\partial_i Q_j, \qquad (3.39)$$

to write the energy density as

$$\mathcal{E} = \frac{1}{2m}tr\left((D_-\psi)^\dagger D_-\psi\right) \qquad (3.40)$$

The energy density is therefore minimized by solutions of the *non-relativistic self-dual Chern-Simons equations* :

$$D_-\psi = 0 \qquad (3.41a)$$

$$F_{+-} = \frac{2}{\kappa}[\psi,\psi^\dagger] \qquad (3.41b)$$

where we recall that $F_{+-} \equiv -2iF_{12}$. Since the self-dual solutions minimize the Hamiltonian density, they provide *static* solutions to the Euler-Lagrange equations of motion (3.32a,3.32b). Alternatively, one can see this directly from inspection of the static equations of motion. The nonabelian version of the factorization identity (2.27) reads

57

$$\vec{D}^2 \psi \equiv D_+ D_- \psi + i[F_{12}, \psi] \tag{3.42}$$

This identity reduces the matter equation of motion (3.32a) to

$$i D_0 \psi = -\frac{1}{2m} D_+ D_- \psi - \frac{1}{2m\kappa}[[\psi, \psi^\dagger], \psi] \tag{3.43}$$

which, with the self-dual ansatz (3.41a), is solved by static solutions

$$\partial_0 \psi = 0 \tag{3.44}$$

with

$$A_0 = \frac{i}{2m\kappa}[\psi, \psi^\dagger] \tag{3.45}$$

Note that this solution is consistent with the electric field equation of motion because the self-dual current takes the simple form

$$J^+ \equiv J^1 + iJ^2 = -\frac{i}{2m}[\psi^\dagger, D_+ \psi] \tag{3.46}$$

An important property of the adjoint coupling nonrelativistic self-dual Chern-Simons equations (3.41) is that they can be obtained by dimensional reduction of the four dimensional self-dual Yang-Mills equations for a nonabelian gauge theory [103,58]. The signature $(2, 2)$ self-dual Yang-Mills equations are

$$F_{12} = F_{34} \tag{3.47a}$$

$$F_{13} = F_{24} \tag{3.47b}$$

$$F_{14} = -F_{23} \tag{3.47c}$$

Taking all fields to be independent of x^3 and x^4, these reduce to

$$F_{12} = [A_3, A_4] \tag{3.48a}$$

$$D_1 A_3 = D_2 A_4 \qquad\qquad (3.48\text{b})$$

$$D_1 A_4 = -D_2 A_3 \qquad\qquad (3.48\text{c})$$

Making the identification

$$\psi = \sqrt{\frac{\kappa}{2}} \, (A_3 - iA_4)$$

$$\psi^\dagger = -\sqrt{\frac{\kappa}{2}} \, (A_3 + iA_4) \qquad\qquad (3.49)$$

the equations (3.48) become the nonrelativistic self-dual Chern-Simons equations (3.41). These dimensionally reduced self-dual Yang-Mills equations have been studied in their own right, independent of any connection with Chern-Simons theories, in the mathematical literature [109,52].

C. Algebraic Ansatze and Toda Theories

Before classifying the general solutions to the nonrelativistic self-dual Chern-Simons equations (3.41), it is instructive to consider certain special cases in which simplifying algebraic *Ansatze* for the fields reduce these equations to familiar integrable nonlinear equations [58]. Such ansatze may also be studied for the general matter coupling case in Section III A, but the most natural case is the adjoint coupling described in Section III B. In this case the matter field ψ and the gauge fields A_μ take values in the same arbitrary representation of the gauge algebra.

First, choose the fields to have the following Lie algebra decomposition

$$A_\mu = i \sum_{a=1}^{r} A_\mu^a H_a \qquad\qquad (3.50\text{a})$$

$$\psi = \sum_{a=1}^{r} \psi^a E_a \qquad (3.50b)$$

Here, H_a refers to the Cartan subalgebra generators and E_a to the simple root step operator generators of the gauge Lie algebra, normalized according to a Chevalley basis [117,32]. (For ease of presentation we consider only simply-laced algebras here, and to be particularly explicit we concentrate our attention on $SU(N)$. Other algebras require the inclusion of additional index factors, as is familiar from Toda theories [158,40,89].) The Chevalley basis is defined by the commutators and traces

$$[H_a, H_b] = 0$$

$$[E_a, E_{-b}] = \delta_{ab} H_a$$

$$[H_a, E_{\pm b}] = \pm C_{ab} E_{\pm b} \qquad (3.51a)$$

$$tr\,(E_a E_{-b}) = \delta_{ab}$$

$$tr\,(H_a H_b) = C_{ab}$$

$$tr\,(H_a E_{\pm b}) = 0 \qquad (3.51b)$$

The indices a and b run over $1 \ldots r$, where r is the rank of the gauge algebra $\mathcal{G}$, and $E_{-a} = E_a^T$. The $r \times r$ matrix C_{ab} is the Cartan matrix of $\mathcal{G}$, which expresses the inner products of the simple roots $\vec{\alpha}^{(a)}$:

$$C_{ab} = \vec{\alpha}^{(a)} \cdot \vec{\alpha}^{(b)} \qquad (3.52)$$

where we have normalized all the root vectors to have $\vec{\alpha}^2 = 2$ (such a normalization is possible for simply-laced algebras).

For example, for $SU(N)$, which has rank $N - 1$, the Cartan matrix C is the $(N - 1) \times (N - 1)$ symmetric tridiagonal matrix (familiar from the theory of numerical analysis):

$$C = \begin{pmatrix} 2 & -1 & 0 & \cdots & & 0 \\ -1 & 2 & -1 & 0 & & \\ 0 & -1 & 2 & -1 & 0 & \\ \vdots & & & & & \vdots \\ 0 & \cdots & & 0 & -1 & 2 \end{pmatrix} \qquad (3.53)$$

In the defining representation of $SU(N)$, the Chevalley basis Cartan subalgebra generators H_a and simple-root step operators E_a may be taken as the $N \times N$ matrices

$$(H_a)_{ij} = \delta_{ai}\delta_{aj} - \delta_{a+1,i}\delta_{a+1,j} \qquad\qquad a = 1 \ldots (N - 1) \quad (3.54a)$$

$$(E_a)_{ij} = \delta_{ai}\delta_{a+1,j} \qquad\qquad a = 1 \ldots (N - 1) \qquad (3.54b)$$

It is a simple matter to check that these matrices satisfy the Chevalley basis commutation relations and trace relations in (3.51) with the Cartan matrix in (3.53). For the defining representation of $SU(N)$, the ansatz (3.50) expresses the fields ψ and A_μ as the following $N \times N$ matrix-valued fields:

$$\psi = \begin{pmatrix} 0 & \psi^1 & 0 & \cdots & & 0 \\ 0 & 0 & \psi^2 & \cdots & & 0 \\ & & \ddots & & & \\ & & & & 0 & \psi^{N-1} \\ 0 & & \cdots & & & 0 \end{pmatrix} \qquad (3.55a)$$

$$A_\mu = i \begin{pmatrix} A_\mu^1 & 0 & 0 & \cdots & 0 \\ 0 & A_\mu^2 - A_\mu^1 & 0 & \cdots & 0 \\ & & \ddots & & \\ 0 & & \cdots & & -A_\mu^{N-1} \end{pmatrix} \qquad (3.55b)$$

Note that the ansatz (3.50) is *algebraically consistent* with the self-duality equations (3.41) in the sense that both F_{+-} and $[\psi, \psi^\dagger]$ lie only in the Cartan subalgebra, and both $D_-\psi$ and $[A_-, \psi]$ lie only in the generators corresponding to the simple root step operators. With this *ansatz* for the fields, the first of the nonrelativistic self-dual Chern-Simons equations (3.41a) reduces to the set of equations

$$\partial_- ln\psi_a = -i \sum_{b=1}^{r} C_{ab} A_-^b \qquad (3.56)$$

When combined with its adjoint, and with the other nonrelativistic self-dual Chern-Simons equation (3.41b), we find the classical Toda equations

$$\nabla^2 ln\rho_a = -\frac{2}{\kappa} \sum_{b=1}^{r} C_{ab}\, \rho_b \qquad (3.57)$$

where

$$\rho_a \equiv |\psi^a|^2 \qquad (3.58)$$

For $SU(2)$, $r = 1$ and the Cartan matrix is just the single number 2, so the Toda equations (3.57) reduce to the Liouville equation

$$\nabla^2 ln\rho = -\frac{4}{\kappa}\rho \qquad (3.59)$$

which Liouville showed to be integrable and indeed "solvable" [182] - in the sense that the general real positive and regular solution can be expressed in terms of a single holomorphic function $f = f(x^-)$ as in (2.79)

$$\rho = \frac{\kappa}{2}\nabla^2 ln\left(1 + f(x^-)\bar{f}(x^+)\right) \qquad (3.60)$$

Kostant [158], and Leznov and Saveliev [177,178] have shown that the classical Toda equations (3.57) are similarly integrable (and indeed solvable), with the general real solutions for ρ_a being expressible in terms of r arbitrary holomorphic functions, where r is the rank of the algebra. For $SU(N)$ it is possible to adapt the Kostant-Leznov-Saveliev solutions to a simple form reminiscent of the Liouville solution (3.60):

$$\rho_a = \frac{\kappa}{2}\nabla^2 ln\, det\left(M_a^\dagger(x^+)M_a(x^-)\right) \qquad a = 1\ldots(N-1)$$

$$(3.61)$$

where M_a is the $N\times a$ *rectangular* matrix

$$M_a = (u\,\,\partial_- u\,\,\partial_-^2 u\,\,\ldots\,\,\partial_-^{a-1}u) \qquad (3.62)$$

with u being an N-component column vector containing $N-1$ arbitrary holomorphic functions $f_1(x^-)$, $f_2(x^-)$, ..., $f_{N-1}(x^-)$:

$$u = \begin{pmatrix} 1 \\ f_1(x^-) \\ f_2(x^-) \\ \vdots \\ f_{N-1}(x^-) \end{pmatrix} \qquad (3.63)$$

Now define the determinants

$$W_a = det\left(M_a^\dagger M_a\right) \qquad (3.64)$$

For example,

$$W_1 = 1 + |f_1|^2 + \ldots + |f_{N-1}|^2 \qquad (3.65)$$

$$W_2 = det \begin{pmatrix} (1 + |f_1|^2 + \ldots + |f_{N-1}|^2) & (f_1' \bar{f}_1 + \ldots f_{N-1}' \bar{f}_{N-1}) \\ (f_1 \bar{f}_1' + \ldots f_{N-1} \bar{f}_{N-1}') & (f_1' \bar{f}_1' + \ldots f_{N-1}' \bar{f}_{N-1}') \end{pmatrix}$$

$$(3.66)$$

These Wronskian-like determinants satisfy the following remarkable nonlinear relation

$$W_a \partial_+ \partial_- W_a - \partial_+ W_a \partial_- W_a = W_{a-1} W_{a+1} \qquad (3.67)$$

with $W_0 \equiv 1$. Furthermore, this relation truncates naturally as

$$W_N \partial_+ \partial_- W_N - \partial_+ W_N \partial_- W_N \equiv W_N^2 \partial_+ \partial_- ln W_N = 0 \qquad (3.68)$$

This truncation occurs because W_N factorizes

$$W_N \equiv det \left(M_N^\dagger M_N \right) = det \left(M_N^\dagger \right) det \left(M_N \right) \qquad (3.69)$$

which means that $ln W_N$ separates into a sum of a function of x^- and a function of x^+:

$$ln W_N = ln det \left(M_N^\dagger (x^+) \right) + ln det \left(M_N (x^-) \right) \qquad (3.70)$$

Note that this factorization only occurs for W_N since M_N and $M_N^\dagger$ are *square* matrices, whereas the M_a and $M_a^\dagger$ for $a < N$ are *rectangular* matrices.

Radially symmetric solutions, analogous to the radially symmetric solutions to the Liouville equation discussed in Section II E, may be obtained by choosing the holomorphic functions $f_a(x^-)$ appearing in (3.63) to be

$$f_a(x^-) = \left(\frac{x^-}{x_a^-} \right)^{-n_a} = \left(\frac{r}{r_a} \right)^{-n_a} e^{in_a\theta} \qquad (3.71)$$

where the r_a ($a = 1 \ldots N - 1$) are constant characteristic length scales, and for single-valued solutions the powers n_a must be integers. The resulting charge densities are

$$\rho_a = \frac{\kappa}{2}\nabla^2 ln\left(1 + p_a(r^2)\right) \tag{3.72}$$

where the $p_a(r^2)$ are polynomials in r^2. The associated charges are

$$\mathcal{Q}_a = \pi\kappa \int_0^\infty r\, dr \nabla^2 ln\left(1 + p_a(r^2)\right)$$

$$= 2\pi\kappa n_{\text{max}}^a \tag{3.73}$$

where n_{max}^a is the maximum power of r^2 in $p_a(r^2)$.

An alternative, extended, *ansatz* for the fields involves the matter field choice

$$\psi = \sum_{a=1}^r \psi^a E_a + \psi^M E_{-M} \tag{3.74}$$

while the gauge field still has its diagonal form as in (3.50a). Here E_{-M} is the step operator corresponding to minus the maximal root, which has the following properties (as before, we present them here for $SU(N)$ only - for other algebras there are additional factors which require more notation, but the generalization is clear given the relation to affine Toda theories)

$$[E_a, E_M] = 0 \qquad a = 1 \ldots r \tag{3.75a}$$

$$[H_a, E_{\pm M}] = \pm \left(\sum_b C_{ab}\right) E_{\pm M} \tag{3.75b}$$

$$[E_M, E_{-M}] = \sum_{a=1}^r H_a \tag{3.75c}$$

For example, for the defining representation of $SU(N)$, the ansatz (3.74) for the matter field ψ is

$$\psi = \begin{pmatrix} 0 & \psi^1 & 0 & \cdots & 0 \\ 0 & 0 & \psi^2 & \cdots & 0 \\ & & \ddots & & \\ & & \cdots & 0 & \psi^{N-1} \\ \psi^M & & \cdots & & 0 \end{pmatrix} \tag{3.76}$$

Thus, we see that $[\psi, \psi^\dagger]$ is once again diagonal, so that the self-duality equation (3.41b) is algebraically consistent. Furthermore, the other self-duality equation (3.41a) is algebraically consistent since both $\partial_-\psi$ and $[A_-, \psi]$ have a decomposition in terms of E_a and E_{-M}.

The self-duality equation $D_-\psi = 0$ then becomes

$$\partial_- ln\psi^a = -i \sum_b C_{ab} A^b_- \tag{3.77a}$$

$$\partial_- ln\psi^M = i \sum_{a,b} C_{ab} A^b_- \tag{3.77b}$$

This implies that

$$\partial_- ln\psi^M = \sum_a \partial_- ln\psi^a \tag{3.78}$$

so that ψ^M is not independent of the other fields:

$$\psi^M = \frac{1}{\prod_a \psi^a} \tag{3.79}$$

(where we have made a rescaling to remove an arbitrary function of x^+). The other self-duality equation involves the charge density

$$[\psi, \psi^\dagger] = \sum_a \rho^a H_a + \rho^M \sum_a H_a \tag{3.80}$$

so that combining with (3.77) we arrive at the equations

66

$$\partial_+\partial_- ln\rho^a = -\frac{2}{\kappa}\left(\sum_b C_{ab}\rho^a - \rho^M\left(\sum_b C_{ab}\right)\right) \tag{3.81}$$

Note that $\sum_b C_{ab} = 1$ for $a = 1$ or $a = N - 1$, and is zero otherwise. For example, for $SU(2)$ (3.81) becomes the Sinh-Gordon equation

$$\partial_+\partial_- ln\rho = -\frac{4}{\kappa}\left(\rho - \frac{1}{\rho}\right) \tag{3.82}$$

The equations (3.81) may be re-expressed as the *affine* Toda equations

$$\partial_+\partial_- ln\rho_a = -\frac{2}{\kappa}\sum_{b=1}^{r+1}\tilde{C}_{ab}\rho_b \tag{3.83}$$

where $\tilde{C}$ is the $(r + 1) \times (r + 1)$ affine Cartan matrix, and $\rho_{r+1} \equiv \rho^M$. The affine Cartan matrix $\tilde{C}$ is a singular matrix obtained by extending the classical Cartan matrix (3.52). Since the extended Cartan matrix $\tilde{C}$ is singular, one of the densities may be eliminated in terms of the others, as in (3.79), and the remaining equations become (3.81). For $SU(2)$,

$$\tilde{C} = \begin{pmatrix} 2 & -2 \\ -2 & 2 \end{pmatrix} \tag{3.84}$$

while for $SU(N)$ with $N \geq 3$, $\tilde{C}$ is the $N \times N$ symmetric matrix

$$\tilde{C} = \begin{pmatrix} 2 & -1 & 0 & \cdots & & -1 \\ -1 & 2 & -1 & 0 & & \\ 0 & -1 & 2 & -1 & 0 & \\ \vdots & & & & & \vdots \\ -1 & \cdots & & 0 & -1 & 2 \end{pmatrix} \tag{3.85}$$

The affine Toda equations (for any classical Lie algebra) are known to be integrable [158,180,181,192], although it is not possible to write simple convergent expressions such as (3.61) for the solutions.

D. 'Yang' Approach

There is another useful way to understand these various algebraic reductions of the nonrelativistic self-dual Chern-Simons equations. In two dimensional space we can (locally) express the gauge field as

$$A_- = G^{-1}\partial_- G \qquad\qquad (3.86a)$$

$$A_+ \equiv -A_-^\dagger = -\partial_+ G^\dagger G^{\dagger-1} \qquad\qquad (3.86b)$$

where G is an element of the complexification of the gauge group (for applications of this representation in other problems see, for example, [252,207,160,25]). Being a complex matrix of unit determinant, G can always be decomposed as

$$G = H\,U \qquad\qquad (3.87)$$

where H is hermitean (and has determinant of magnitude 1) and U is unitary. Then

$$A_- = U^{-1}\partial_- U + U^{-1}\left(H^{-1}\partial_- H\right)U \qquad\qquad (3.88a)$$

$$A_+ = U^{-1}\partial_+ U - U^{-1}\left(\partial_+ H H^{-1}\right)U \qquad\qquad (3.88b)$$

Note that only with the hermitean factor $H = \mathbf{1}$ does (3.86,3.88) correspond to a pure gauge. Gauge transformations on $A_\pm$ correspond to different choices of the unitary factor U. The field strength corresponding to (3.86,3.88) has the following simple form, with the unitary factors factoring out as expected:

$$F_{+-} = U^{-1}\left(H\partial_+\left(H^{-2}\partial_- H^2\right)H^{-1}\right)U \qquad\qquad (3.89)$$

With the gauge field represented as in (3.86), the solution to the first self-duality equation $D_-\psi = 0$ is trivially:

$$\psi = G^{-1}\psi_0(x^+)G$$

$$= U^{-1}\left(H^{-1}\psi_0(x^+)H\right)U \qquad (3.90)$$

for *any* $\psi_0(x^+)$. Then the charge density is

$$[\psi, \psi^\dagger] = U^{-1}[H^{-1}\psi_0 H, H\psi_0^\dagger H^{-1}]U \qquad (3.91)$$

Combining this expression with the formula (3.89) for the gauge curvature, we find that the second self-duality equation (3.41b) becomes the following gauge invariant equation for H^2:

$$\partial_+\left(H^{-2}\partial_- H^2\right) = \frac{2}{\kappa}\left(\psi_0^\dagger H^{-2}\psi_0 H^2 - H^{-2}\psi_0 H^2 \psi_0^\dagger\right) \qquad (3.92)$$

Thus far, no special choices have been made and equation (3.92) is still completely general. Now, if we choose to write H^2 as

$$H^2 = e^{-\Phi} \qquad (3.93)$$

where Φ is restricted to the Cartan subalgebra, then (3.92) simplifies to

$$\partial_+\partial_-\Phi = \frac{2}{\kappa}\left(\psi_0^\dagger e^{-\Phi}\psi_0 e^{\Phi} - e^{-\Phi}\psi_0 e^{\Phi}\psi_0^\dagger\right) \qquad (3.94)$$

These equations follow as equations of motion from the two-dimensional Euclidean Lagrange density

$$\mathcal{L} = tr\left(\frac{1}{2}\partial_\mu\Phi\partial_\mu\Phi - \frac{2}{\kappa}e^{\Phi}\psi_0 e^{-\Phi}\psi_0^\dagger\right) \qquad (3.95)$$

For suitable choices of $\psi_0(x^+)$ as a *constant* field this Lagrange density becomes the (Euclidean) classical Toda or affine Toda Lagrange density. For example, for $SU(N)$, with ψ_0 chosen as

$$\psi_0 = \sum_a E_a \tag{3.96}$$

the Lagrange density (3.95) becomes that of the *classical* Toda theory, while if $\psi_0(x^+)$ is chosen to be the constant field

$$\psi_0 = \sum_a E_a + E_{-M} \tag{3.97}$$

then (3.95) becomes the *affine* Toda Lagrange density. This example provides another illustration of the connection between the self-dual Chern-Simons equations and the Toda equations.

E. Self-Duality Equations and the Chiral Model

Having considered some special cases in which the nonrelativistic self-dual Chern-Simons equations reduce to well-known integrable equations in two-dimensional Euclidean space, we now consider the question of finding the most general solutions [59,61]. The key to the possibility of finding all solutions lies in the fact that there exists a special gauge transformation g which converts the two coupled self-duality equations (3.41) into a *single* equation

$$\partial_-\chi = [\chi^\dagger, \chi] \tag{3.98}$$

where χ is the gauge transformed matter field

$$\chi = \sqrt{\frac{2}{\kappa}} g \psi g^{-1} \tag{3.99}$$

The existence of such a gauge transformation g^{-1} follows from the following zero-curvature formulation of the self-dual Chern-Simons equations [58,59]. Define

$$\mathcal{A}_+ \equiv A_+ - \sqrt{\frac{2}{\kappa}} \psi \tag{3.100a}$$

$$\mathcal{A}_- \equiv A_- + \sqrt{\frac{2}{\kappa}} \psi^\dagger \qquad (3.100b)$$

Then the gauge curvature associated with $\mathcal{A}_\pm$ is

$$\mathcal{F}_{+-} \equiv \partial_+ \mathcal{A}_- - \partial_- \mathcal{A}_+ + [\mathcal{A}_+, \mathcal{A}_-]$$

$$= \partial_+ A_- - \partial_- A_+ + [A_+, A_-] - \frac{2}{\kappa} [\psi, \psi^\dagger]$$

$$+ \sqrt{\frac{2}{\kappa}} D_- \psi + \sqrt{\frac{2}{\kappa}} (D_- \psi)^\dagger \qquad (3.101)$$

As a consequence of the self-duality equations (3.41) we find that

$$\mathcal{F}_{+-} = 0 \qquad (3.102)$$

Therefore, at least locally, one can write $\mathcal{A}_\pm$ as a pure gauge

$$\mathcal{A}_\pm = g^{-1} \partial_\pm g \qquad (3.103)$$

for some g in the gauge group. If χ is defined as the gauge trans-formed field in (3.99) then

$$\partial_- \chi = \sqrt{\frac{2}{\kappa}} g \left([g^{-1} \partial_- g, \psi] + \partial_- \psi \right) g^{-1}$$

$$= \sqrt{\frac{2}{\kappa}} g \left([A_- - \mathcal{A}_-, \psi] \right) g^{-1}$$

$$= \frac{2}{\kappa} g [\psi^\dagger, \psi] g^{-1}$$

$$= [\chi^\dagger, \chi] \qquad (3.104)$$

which is just the single equation (3.98). Alternatively, note that

$$D_- \psi = \sqrt{\frac{\kappa}{2}} g^{-1} \left(\partial_- \chi - [\chi^\dagger, \chi] \right) g \qquad (3.105)$$

and

71

$$\partial_+ A_- - \partial_- A_+ + [A_+, A_-] - \frac{2}{\kappa}[\psi, \psi^\dagger]$$

$$= g^{-1}\left(\partial_- \chi + \partial_+ \chi^\dagger - 2[\chi^\dagger, \chi]\right)g \qquad (3.106)$$

The equation (3.98) can be converted into the chiral model equation

$$\partial_+(h^{-1}\partial_- h) + \partial_-(h^{-1}\partial_+ h) = 0 \qquad (3.107)$$

upon defining

$$\chi \equiv \frac{1}{2}h^{-1}\partial_+ h \qquad (3.108)$$

for some h in the gauge group. With χ as in (3.108), the chiral model equation (3.107) follows because (3.98) implies that

$$\partial_- \chi = \partial_+ \chi^\dagger \qquad (3.109)$$

The factor of $\frac{1}{2}$ is required in (3.108) in order to satisfy the nonlinear equation (3.98):

$$\partial_- \chi = \frac{1}{2}\left(\partial_- \chi + \partial_+ \chi^\dagger\right)$$

$$= \frac{1}{4}\left(-h^{-1}\partial_- h h^{-1}\partial_+ h + h^{-1}\partial_+ h h^{-1}\partial_- h\right)$$

$$= [\chi^\dagger, \chi] \qquad (3.110)$$

Given any solution h of the chiral model equation (3.107), or alternatively any solution χ of (3.98), we automatically obtain a solution of the original nonrelativistic self-dual Chern-Simons equations:

$$\psi^{(0)} = \sqrt{\frac{\kappa}{2}}\chi \qquad\qquad A_+^{(0)} = \chi \qquad\qquad A_-^{(0)} = -\chi^\dagger \qquad (3.111)$$

Any other solution is a gauge transform of this solution.

The chiral model equations are also known as the "harmonic map equations" [235,244] because if we regard $J_\pm = h^{-1}\partial_\pm h$ as a connection, then it satisfies both

$$\partial_+ J_- + \partial_- J_+ = 0$$

$$\partial_+ J_- - \partial_- J_+ + [J_+, J_-] = 0 \qquad (3.112)$$

and so has zero divergence and zero curl.

So far, all our discussion has been in terms of *local* analysis. The *global* condition which permits the classification of solutions to the chiral model equation (3.107) is the condition of *finiteness* of the chiral model "action functional" (also referred to in the literature as the "energy functional")

$$\mathcal{E}[h] \equiv -\frac{1}{2} \int d^2x \; tr(h^{-1}\partial_- h h^{-1}\partial_+ h) \qquad (3.113)$$

This finiteness condition has direct physical relevance in the related nonrelativistic Chern-Simons system because we see from the definition (3.108)

$$\mathcal{E}[h] = 2 \int d^2x \; tr(\chi\chi^\dagger)$$

$$= \frac{4}{\kappa} \int d^2x \; tr(\psi\psi^\dagger)$$

$$= \frac{4}{\kappa} Q \qquad (3.114)$$

where Q is the conserved gauge invariant matter charge in (3.35a). Thus, the *finite action* solutions of the chiral model equation (3.107) correspond precisely to the *finite charge* solutions of the nonrelativistic self-dual Chern-Simons equations (3.41).

In addition to being physically significant, this finiteness condition is mathematically crucial because it permits the chiral model

73

solutions on $\mathbf{R}^2$ to be classified by conformal compactification to the sphere S^2 [235,244]. Indeed, Uhlenbeck has classified all finite action chiral model solutions for $SU(N)$ in terms of "uniton" factors (which will be discussed in Section III F).

F. Unitons and General Solutions to the SDCS Equations

Before discussing the *general* classification of finite charge solutions, we introduce the simplest such solutions, the "single unitons", upon which the general solutions are constructed [235]. A "single uniton" solution, h, of the $SU(N)$ chiral model equation (3.107) has the form

$$h = 2p - \mathbf{1} \qquad (3.115)$$

where p is a "holomorphic projector" satisfying the following three conditions:

$$p^\dagger = p \qquad (3.116a)$$

$$p^2 = p \qquad (3.116b)$$

$$(\mathbf{1} - p)\partial_+ p = 0 \qquad (3.116c)$$

Note that $h \in SU(N)$ because

$$h^{-1} = 2p - \mathbf{1} = h^\dagger \qquad (3.117)$$

These *single* uniton solutions (3.115,3.116) are fundamental to the chiral model system because as a consequence of the conditions (3.116) we find that

$$h^{-1}\partial_\pm h = \pm 2\partial_\pm p \qquad (3.118)$$

74

From this it immediately follows that h satisfies the chiral model equation (3.107). These single uniton solutions are *also* solutions of the CP^{N-1} model since h satisfies the additional CP^{N-1} condition, $h^2 = 1$, as a result of p being a projector.

In terms of the χ field defined in (3.108), the single uniton solutions take the simple form

$$\chi = \partial_+ p \qquad (3.119)$$

We can check that this χ does in fact satisfy the equation (3.98) because

$$\partial_- \chi = \partial_- \partial_+ p$$

$$= \partial_+ \left(\partial_- p p \right)$$

$$= \partial_+ \partial_- p p + \partial_- p \partial_+ p$$

$$= [\chi^\dagger, \chi] \qquad (3.120)$$

which is just the equation (3.98). Therefore (3.119) gives a solution to the nonrelativistic self-dual Chern-Simons equations as in (3.111).

The general holomorphic projector satisfying the conditions (3.116) can be expressed as

$$p = M \left(M^\dagger M \right)^{-1} M^\dagger \qquad (3.121)$$

where M is any (rectangular) matrix such that $M = M(x^-)$ [235]. It is easy to see that such an M is a *hermitean* projector. The third condition, (3.116c), is equivalent to $\partial_+ p \, p = 0$, which follows immediately from the fact that

$$\partial_+ p = M \left(M^\dagger M \right)^{-1} \partial_+ M^\dagger \left(1 - p \right) \qquad (3.122)$$

75

The next step towards the construction of general solutions involves the process of "composing" uniton solutions, as follows. Suppose $h_1 = 2p_1 - \mathbf{1}$ is a single uniton solution with p_1 satisfying the conditions (3.116) for a holomorphic projector. Further, let $h_2 = 2p_2 - \mathbf{1}$ be such that $p_2 = p_2^\dagger$ and $p_2^2 = p_2$. Then

$$h = h_1 h_2 \tag{3.123}$$

is a solution of the chiral model equation (3.107) provided the following first-order algebro-differential conditions are met:

$$(i) \qquad (\mathbf{1} - p_2) \left(\partial_+ + \frac{1}{2} h_1^{-1} \partial_+ h_1 \right) p_2 = 0 \tag{3.124a}$$

$$(ii) \qquad (\mathbf{1} - p_2) \left(\frac{1}{2} h_1^{-1} \partial_- h_1 \right) p_2 = 0 \tag{3.124b}$$

Given these conditions (3.124), the composite uniton h in (3.123) satisfies

$$h^{-1} \partial_\pm h = \pm 2 \left(\partial_\pm p_1 + \partial_\pm p_2 \right) \tag{3.125}$$

and so, once again, the chiral model equation (3.107) is immediately solved.

This procedure of composing uniton-type solutions can be continued, but since the p matrices involved are projectors, there is a limit to how many independent projections can be made. For $SU(N)$, at most $N-1$ such terms can be combined in this manner, as expressed in Uhlenbeck's theorem:

THEOREM (K. Uhlenbeck [235]; see also J. C. Wood [251]): *Every finite action solution h of the $SU(N)$ chiral model equation (3.107) may be uniquely factorized as a product of "uniton" factors*

$$h = \pm h_0 \prod_{i=1}^m (2p_i - 1) \tag{3.126}$$

where:

a) $h_0 \in SU(N)$ *is constant;*

b) each p_i *is a Hermitean projector* $(p_i^\dagger = p_i$ *and* $p_i^2 = p_i)$*;*

c) defining $h_j = h_0 \prod_{i=1}^{j}(2p_i - 1)$*, the following linear relations must hold:*

$$(1 - p_i) \left(\partial_+ + \frac{1}{2} h_{i-1}^{-1} \partial_+ h_{i-1} \right) p_i = 0$$

$$(1 - p_i) \, h_{i-1}^{-1} \partial_- h_{i-1} \, p_i = 0 \qquad (3.127)$$

d) $m \leq N - 1$.

The $\pm$ sign in (3.126) has been inserted to allow for the fact that Uhlenbeck and Wood considered the gauge group $U(N)$ rather than $SU(N)$.

An important implication of this theorem is that for $SU(2)$ *all* finite action solutions of the chiral model have the "single uniton" form

$$h = (2p - 1) \qquad (3.128)$$

with p being a holomorphic projector satisfying the conditions (3.116). These $SU(2)$ single uniton solutions are essentially the CP^1 model solutions [256] because a single uniton satisfies the addition CP^1 condition

$$h^2 = 1 \qquad (3.129)$$

For $SU(N)$ with $N \geq 3$ one must consider composite unitons in addition to the single unitons. Thus the Toda ansatz solutions discussed in Section III C are not sufficient to describe all solutions to the self-duality equations (although they do provide the 'building blocks' for the general solutions). This is consistent with the results of index theory analyses counting the number of parameters for solutions of the self-duality equations [254]. It becomes increasingly

77

difficult to give a simple characterization of all possible projectors satisfying the subsidiary linear conditions (3.127) specified in Uhlenbeck's construction. However, Wood has presented a systematic parametrization of these higher unitons, for any $SU(N)$, in terms of a sequence of projectors into Grassmannian factors [251]. A detailed analysis of the $SU(3)$ and $SU(4)$ cases is also given in [203,204].

Another useful result from the mathematical literature on the chiral model equation (3.107) is that the chiral model energy (3.113) is *quantized* in integral multiples of 8π [237]. Becuase of the correspondence (3.111,3.108) between chiral model solutions and solutions of the self-duality equations (3.41), this implies that the abelian Chern-Simons charge $\mathcal{Q} \equiv \int tr(\psi^\dagger \psi)$ is quantized in integral multiples of $2\pi\kappa$.

$$\mathcal{Q} \equiv \int tr(\psi^\dagger \psi) = \frac{\kappa}{2} \int tr(\chi^\dagger \chi) = \frac{\kappa}{4}\mathcal{E}[h] = 2\pi\kappa n \qquad (3.130)$$

This quantization condition is consistent with the quantized non-abelian charges (3.73) found for the particular $SU(N)$ Toda solutions in Section III C, because for the Toda solutions $\mathcal{Q} = \sum_a \mathcal{Q}_a$, and each $\mathcal{Q}_a$ is similarly quantized.

G. Unitons and Toda Theories

Given that the uniton construction described in Section III F gives *all* finite charge solutions to the self-dual Chern-Simons equations (3.41), it is important to ask how these multi-uniton solutions to the chiral model equations are related to the explicit Toda-type solutions discussed previously in (3.57-3.61). At first sight, the solutions (3.61) and (3.126) look completely different, but this is because they correspond to different gauges. Notice that while the algebraic *Ansätze* (3.50,3.74) each leads to a non-Abelian charge density

$\rho = [\psi, \psi^\dagger]$ which is *diagonal*, the chiral model solutions (3.111) have charge density $\rho^{(0)} = \frac{\kappa}{2}[\chi, \chi^\dagger]$ which need not be diagonal. However, $\rho^{(0)}$ is always hermitean, and so it can be diagonalized by a gauge transformation. It is still a nontrivial algebraic task to implement this diagonalization explicitly, but this can be done for the $SU(N)$ solutions, revealing an interesting new link between the chiral model and the Toda system [59,91,92].

It is instructive to illustrate this procedure with the $SU(2)$ case first. Here, Uhlenbeck's theorem implies that the only finite charge solution has the form $\chi = \partial_+ p$, where p is a holomorphic projector as in (3.121). For $SU(2)$ we can only project onto a *line* in $\mathbf{C}^2$, so we take

$$M(x^-) = \begin{pmatrix} f(x^-) \\ 1 \end{pmatrix} \tag{3.131}$$

Then the corresponding projection matrix (3.121) is

$$p = \frac{1}{1 + f\bar{f}} \begin{pmatrix} f\bar{f} & f \\ \bar{f} & 1 \end{pmatrix} \tag{3.132}$$

and the associated solution $\chi = \partial_+ p$ can be expressed in terms of the single holomorphic function $f(x^-)$:

$$\chi = \partial_+ p = \frac{f\partial_+\bar{f}}{(1 + f\bar{f})^2} \begin{pmatrix} 1 & -f \\ 1/f & -1 \end{pmatrix} \tag{3.133}$$

The corresponding matter density is

$$[\chi, \chi^\dagger] = -\frac{|\partial_- f|^2}{(1 + f\bar{f})^3} \begin{pmatrix} 1 - f\bar{f} & 2f \\ 2\bar{f} & -1 + f\bar{f} \end{pmatrix} \tag{3.134}$$

which may be diagonalized by the unitary matrix

$$g = \frac{1}{\sqrt{1 + f\bar{f}}} \begin{pmatrix} -f & 1 \\ 1 & \bar{f} \end{pmatrix} \tag{3.135}$$

to yield the diagonalized charge density

$$[\psi, \psi^\dagger] = \frac{\kappa}{2} g^{-1} [\chi, \chi^\dagger] g$$

$$= \frac{\kappa}{2} \partial_+ \partial_- \ln\left(1 + |f|^2\right) \begin{pmatrix} 1 & 0 \\ 0 & -1 \end{pmatrix}$$

$$= \frac{\kappa}{2} \partial_+ \partial_- \ln \det(M^\dagger M) \begin{pmatrix} 1 & 0 \\ 0 & -1 \end{pmatrix} \qquad (3.136)$$

In this diagonalized form we recognize Liouville's solution (3.60) to the classical $SU(2)$ Toda equation. It is worth emphasizing that for $SU(2)$ the nonrelativistic self-dual Chern-Simons equations (3.41) can be converted, by suitable algebraic ansatze as discussed in Section III C, into the classical Toda (i.e. Liouville) equation or the affine Toda (i.e. sinh-Gordon) equation. However, the above analysis shows that only the classical Toda case (i.e. Liouville) corresponds to finite charge.

A similar construction is possible for the $SU(N)$ case [59,61]. Specifically, let

$$h = (-1)^{\frac{1}{2}N(N+1)} \prod_{a=1}^{N-1} (2p_a - 1) \qquad (3.137)$$

be a product where each p_a is a holomorphic projector

$$p = M_a \left(M_a^\dagger M_a\right)^{-1} M_a^\dagger \qquad (3.138)$$

onto the a-dimensional subspace spanned by the columns of the $N \times a$ rectangular matrix $M_a(x^-)$ in (3.61,3.63):

$$M_a = \begin{pmatrix} 1 & 0 & \dots & 0 \\ f_1 & \partial_- f_1 & \dots & \partial_-^{(a-1)} f_1 \\ f_2 & \partial_- f_2 & \dots & \partial_-^{(a-1)} f_2 \\ \vdots & \vdots & & \vdots \\ f_{N-1} & \partial_- f_{N-1} & \dots & \partial_-^{(a-1)} f_{N-1} \end{pmatrix} \qquad (3.139)$$

Then the h given by (3.137) is a finite action solution of the $SU(N)$ chiral model equation (3.107) and the corresponding solution (3.111) of the nonrelativistic self-dual Chern-Simons equations is

$$\chi = \sum_{a=1}^{N-1} \partial_+ p_a \qquad (3.140)$$

The corresponding charge density $[\chi, \chi^\dagger]$ is not diagonal, but it may be diagonalized by an $SU(N)$ gauge transformation g. To explicitly construct the diagonalizing matrix g, we use the Gramm-Schmit procedure to construct an orthonormal set of vectors

$$\{e_1, e_2, e_3, \ldots, e_N\} \qquad (3.141)$$

from the vectors

$$\{u, \partial_- u, \partial_-^2 u, \ldots, \partial_-^{N-1} u\} \qquad (3.142)$$

Then

$$p_a = \sum_{b=1}^{a} e_b e_b^\dagger \qquad (3.143)$$

and the diagonalizing matrix g has columns given by the basis vectors (3.141)

$$g = (e_1 \, e_2 \, \ldots e_N) \qquad (3.144)$$

With this matrix g in (3.144),

$$g^{-1} \partial_+ p_a g = \left(e_a^\dagger \partial_+ e_{a+1}\right) E_a \qquad a = 1 \ldots (N-1) \qquad (3.145)$$

where the $N \times N$ matrices E_a are given by (3.54b) as the defining representation of the simple root step operators for $SU(N)$. Then we can use the $SU(N)$ Chevalley basis commutation relations (3.51) to express the diagonalized charge density as

81

$$g^{-1}[\chi, \chi^\dagger]g = \sum_{a=1}^{N-1} \left| e_a^\dagger \partial_+ e_{a+1} \right|^2 H_a \qquad (3.146)$$

where the H_a are the diagonal Cartan subalgebra generators (3.54a).
Finally, we note that

$$\partial_+ \partial_- \ln\det\left(M_a^\dagger M_a\right) = \partial_+ \partial_- tr\ln\left(M_a^\dagger M_a\right)$$

$$= \partial_+ tr\left(\left(M_a^\dagger M_a\right)^{-1} M_a^\dagger \partial_- M_a\right)$$

$$= tr\left(\partial_- p_a \partial_+ p_a\right)$$

$$= \left| e_a^\dagger \partial_+ e_{a+1} \right|^2 \qquad (3.147)$$

Therefore, the diagonalized form of the charge density is

$$[\psi, \psi^\dagger] = \frac{\kappa}{2} g^{-1}[\chi, \chi^\dagger]g = \frac{\kappa}{2} \sum_{a=1}^{N-1} \{\partial_+ \partial_- \log \det(M_a^\dagger M_a)\} H_a \qquad (3.148)$$

from which we recognize the H_a basis coefficients as the solutions
(3.61) to the $SU(N)$ classical Toda equations.

IV. ABELIAN RELATIVISTIC MODEL

In this Chapter we generalize the nonrelativistic self-dual Chern-Simons systems discussed in Chapter 2 to the case where the matter fields possess *relativistic* dynamics [113,127]. It is a unique feature of these Chern-Simons theories that it is still possible to find a factorization of the equations of motion into first-order self-duality equations. The energy is bounded below, with the bound saturated by solutions to the self-duality equations, provided the scalar potential has a particular sixth order form with two degenerate vacua. The self-duality equations are similar in form to those of the Abelian Higgs model, although the self-dual Chern-Simons system supports *charged* vortex solutions, while the abelian Higgs model has neutral vortices. The possibility of finding *charged* vortices in a matter-gauge theory which included a Chern-Simons term (in addition to the usual Maxwell or Yang-Mills term) was originally proposed in [49,201,159,136]. While a radial-ansatz analysis of the resulting field equations indicated the presence of vortex-like solutions, no self-dual structure was present in these models. The pure Chern-Simons limit of these models [138] focusses on long-distance properties, and it was in this limit that the role of self-duality was found [113,127]. While explicit closed-form solutions to the resulting self-dual Chern-Simons equations are not known (as is also the case for the abelian Higgs model), much can be said about possible solutions in terms of their global asymptotic properties. In particular, the self-dual Chern-Simons system has both topological vortices and nontopological solitons [128,129]. Another interesting feature of these models concerns the Chern-Simons Higgs mechanism whereby a massive gauge mode is generated in the broken vacuum, even though there is no Maxwell term present [48]. The self-dual potential leads to particular mass relations in the unbroken and broken phases. This, in turn, is related to the presence of

83

an extended supersymmetry: the relativistic self-dual Chern-Simons Lagrangian is the bosonic part of a Lagrangian with $N = 2$ supersymmetry. Indeed, the requirement of $N = 2$ supersymmetry is enough, in this abelian model, to prescribe the self-dual form of the potential [165]. This is an example of a general feature of (relativistic) self-dual models which relates the Bogomol'nyi energy bound to a topological charge that appears as a central charge in the supersymmetry algebra [248,111,112]. Finally, as in the nonrelativistic case, it is possible to extend these models to include a Maxwell term for the gauge field, provided an additional (neutral) scalar field is introduced which has mass degenerate with the massive gauge excitation in the unbroken vacuum [166].

A. Lagrangian and Hamiltonian Formulation

The abelian Chern-Simons system with nonrelativistic matter (discussed in Chapter 2) can be generalized to the case of *relativistic* matter fields. Remarkably, a related notion of self-duality still exists in these relativistic models [113,127,129].

Consider the Lagrange density

$$\mathcal{L} = \frac{\kappa}{2}\epsilon^{\mu\nu\rho}A_\mu \partial_\nu A_\rho - (D_\mu\phi)^* D^\mu\phi - V(|\phi|) \tag{4.1}$$

where this model is defined in $2 + 1$ dimensional Minkowski spacetime, with metric $g_{\mu\nu} = diag(-1, 1, 1)$. The charged matter field ϕ is a complex scalar field and

$$D_\mu\phi = \partial_\mu\phi + iA_\mu\phi \tag{4.2}$$

The potential $V(|\phi|)$ is chosen to be a general sixth order potential

$$V(|\phi|) = c_0 + c_1|\phi|^2 + c_2|\phi|^4 + c_3|\phi|^6 \tag{4.3}$$

84

where we note that such a potential is power-counting renormalizable in $2+1$ dimensions. We shall see in Section IV B that the self-dual system dictates a specific form for the sixth order potential.

This theory has a local $U(1)$ gauge symmetry, and the Euler-Lagrange equations of motion are

$$D_\mu D^\mu \phi = \frac{\partial V}{\partial \phi^*} \qquad (4.4a)$$

$$F_{\mu\nu} = -\frac{1}{\kappa}\epsilon_{\mu\nu\rho}J^\rho \qquad (4.4b)$$

where J^μ is the relativistic matter current

$$J^\mu = -i\left(\phi^* D^\mu \phi - (D^\mu \phi)^* \phi\right) \qquad (4.5)$$

which is conserved

$$\partial_\mu J^\mu = 0 \qquad (4.6)$$

The energy density corresponding to the Lagrange density (4.1) is

$$\mathcal{E} = |D_0 \phi|^2 + |\vec{D}\phi|^2 + V(|\phi|) \qquad (4.7)$$

supplemented by the Gauss law constraint

$$F_{12} = \frac{1}{\kappa}J^0 = -\frac{i}{\kappa}\left(\phi^* D^0 \phi - \left(D^0 \phi\right)^* \phi\right) \qquad (4.8)$$

As is familiar for Chern-Simons theories (in which the gauge field action consists of *only* a Chern-Simons term, with no Maxwell term), the Chern-Simons term does not contribute to the energy density $\mathcal{E}$ since $\mathcal{L}_{CS}$ is first order in spacetime derivatives.

B. Self-Duality Equations

To obtain self-duality equations analogous to the nonrelativistic self-dual Chern-Simons equations (2.34) we proceed in a similar manner, using the identity (2.35) to write

$$|\vec{D}\phi|^2 = |D_\mp\phi|^2 \pm F_{12}|\phi|^2 \mp \frac{1}{2}\epsilon^{ij}\partial_i J_j \qquad (4.9)$$

Then the Chern-Simons Gauss law constraint (4.8) implies that

$$|\vec{D}\phi|^2 = |D_\mp\phi|^2 \pm \frac{i}{\kappa}|\phi|^2\left(\phi^* D_0\phi - (D_0\phi)^*\phi\right) \qquad (4.10)$$

where, as before, we have dropped a spatial total derivative term. Notice the appearance of the *relativistic* current density J_0 which introduces a $D_0\phi$ dependence into this expression for $|\vec{D}\phi|^2$. In the corresponding nonrelativistic equation, Gauss' law relates F_{12} to the nonrelativistic charge density $\rho \equiv |\psi|^2$, and so the corresponding expression for $|\vec{D}\phi|^2$ involves $|D_\mp\phi|^2$ and a purely potential term (see (2.36)).

This $D_0\phi$ dependence may be absorbed into the $|D_0\phi|^2$ term in the energy density $\mathcal{E}$ by writing

$$|D_0\phi|^2 = \left|D_0\phi \mp \frac{i}{\kappa}|\phi|^2\phi\right|^2 \mp \frac{i}{\kappa}|\phi|^2\left(\phi^* D_0\phi - (D_0\phi)^*\phi\right) - \frac{1}{\kappa^2}|\phi|^6$$

$$(4.11)$$

Then the energy density (4.7) may be written as

$$\mathcal{E} = \left|D_0\phi \mp \frac{i}{\kappa}|\phi|^2\phi\right|^2 + |D_\mp\phi|^2 + V(|\phi|) - \frac{1}{\kappa^2}|\phi|^6 \qquad (4.12)$$

Therefore, choosing the scalar potential V to be the purely sextic potential

$$V(|\phi|) = \frac{1}{\kappa^2}|\phi|^6 \qquad (4.13)$$

we would have an expression for $\mathcal{E}$ as a sum of two squares. However, this purely 6th order potential in (4.13) has no mass term and will prove unsuitable, for several reasons to be mentioned below. Rather, it is better to explicitly introduce a mass scale v^2 by writing (instead of (4.11))

$$|D_0\phi|^2 = \left|D_0\phi \mp \frac{i}{\kappa}\left(|\phi|^2 - v^2\right)\phi\right|^2 - \frac{1}{\kappa^2}|\phi|^2\left(|\phi|^2 - v^2\right)^2$$

$$\mp \frac{i}{\kappa}\left(|\phi|^2 - v^2\right)\left(\phi^* D_0\phi - (D_0\phi)^*\phi\right) \qquad (4.14)$$

Then, combining (4.10) and (4.14) we can express the energy density as

$$\mathcal{E} = \left|D_0\phi \mp \frac{i}{\kappa}\left(|\phi|^2 - v^2\right)\phi\right|^2 + |D_{\mp}\phi|^2 \pm \frac{v^2}{\kappa}J^0 \qquad (4.15)$$

where we have now chosen the self-dual potential to be

$$V(|\phi|) = \frac{1}{\kappa^2}|\phi|^2\left(|\phi|^2 - v^2\right)^2 \qquad (4.16)$$

Notice that this approach selects both the *form* of the potential and the overall *magnitude*, which is correlated with the strength κ of the Chern-Simons coupling. This form of the potential has two distinct minima, and they are degenerate. We shall understand the significance of this in more detail later when we come to consider the associated supersymmetry.

From (4.15) we see that the energy is bounded below by a constant proportional to the net magnetic flux

$$E \geq \pm\frac{v^2}{\kappa}\int d^2x\, J^0 = v^2\int d^2x\, B \qquad (4.17)$$

To have a positive lower bound for the energy (with positive charge $Q^0 \equiv \int J^0$) we choose κ to be positive and we choose the upper sign of $\pm$. The lower bound (4.17) on the energy is saturated when the fields satisfy the *first order* equations

87

$$D_-\phi = 0 \tag{4.18a}$$

$$D_0\phi = \frac{i}{\kappa}\left(|\phi|^2 - v^2\right)\phi \tag{4.18b}$$

There is also the Gauss law constraint to be satisfied, and this may be combined with (4.18b) to yield the "relativistic self-dual Chern-Simons equations":

$$D_-\phi = 0 \tag{4.19a}$$

$$
\begin{aligned}
F_{12} &= -\frac{2}{\kappa^2}|\phi|^4 + \frac{2v^2}{\kappa^2}|\phi|^2 \\
&= -\frac{2}{\kappa^2}|\phi|^2\left(|\phi|^2 - v^2\right)
\end{aligned}
\tag{4.19b}
$$

These self-duality equations should be compared with the nonrelativistic self-dual Chern- Simons equations (2.34). They are first order in derivatives, and the first equations are the same in each case. The second equations, (2.34b) and (4.19b), are of the same form but the relativistic equation has an additional quartic term in the magnitude of the scalar field. In Section IV D we shall consider in detail the nonrelativistic limit, in which the relativistic self-dual Chern-Simons equations (4.19) reduce to the nonrelativistic self-dual Chern- Simons equations (2.34). It is also important to note that the relativistic self-dual Chern-Simons equations (4.19) are similar in form to the Bogomol'nyi equations (1.9) of the abelian Higgs model.

Note that the relativistic self-dual solutions are not necessarily static, although their time dependence is trivial. In fact, the most natural resolution of the first order equation (4.18b) is to give the scalar field ϕ a simple time dependence

$$\phi = e^{-imx^0}\phi_{\text{static}} \tag{4.20}$$

88

corresponding to the rest mass ($m = v^2/\kappa$) energy, and to identify the time component of the gauge field as

$$A_0 = \frac{1}{\kappa}|\phi|^2 \tag{4.21}$$

just as in the nonrelativistic case (2.31).

However, unlike their nonrelativistic counterparts (2.34), the relativistic self-dual Chern-Simons equations (4.19) cannot be solved in closed form. This lack of explicit exact solutions is also a feature of the abelian Higgs model [196,24,137]. Nevertheless, as for the abelian Higgs model, much can be deduced about the solutions from their asymptotic and global behavior. To investigate the properties of solutions we decompose the complex scalar field as

$$\phi = e^{-i\omega}|\phi| \tag{4.22}$$

Then the first self-duality equation (4.19a) implies that the corresponding vector potential is

$$A_i = \partial_i\omega + \epsilon_{ij}\partial_j ln|\phi| \tag{4.23}$$

while the magnitude $|\phi|$ of the matter field satisfies the nonlinear equation

$$\nabla^2 ln|\phi|^2 = \frac{4}{\kappa^2}|\phi|^2 \left(|\phi|^2 - v^2\right) \tag{4.24}$$

If v^2 were zero, corresponding to the potential (4.13) then this equation would reduce to

$$\nabla^2 ln|\phi|^2 = \frac{4}{\kappa^2}|\phi|^4 \tag{4.25}$$

which has no real positive and regular solutions for $|\phi|^2$. This is one reason for having introduced the mass term in the self-dual potential (4.16).

In Section IV D we shall show that in the nonrelativistic limit the equation (4.24) reduces to the Liouville equation (2.77) which is a completely solvable nonlinear equation. However, in the *relativistic* theory we must be content with approximate solutions, for which we can investigate asymptotic and global properties, as is discussed in Section IV C.

C. Self-Dual Solutions: Topological Vortices and

Nontopological Solitons

In order to obtain *finite* energy solutions to the Euler-Lagrange equations of motion (4.4), the matter field configuration ϕ must tend at spatial infinity to a minimum of the potential $V(|\phi|)$. For the self-dual potential (4.16) there are two inequivalent minima, and corrspondingly there are two different types of solutions to the nonlinear equation (4.24):

$$|\phi|^2 \to v^2, \qquad\qquad r \to \infty \qquad\qquad (4.26\text{a})$$

$$|\phi|^2 \to 0, \qquad\qquad r \to \infty \qquad\qquad (4.26\text{b})$$

The former case corresponds to topologically stable solutions with broken symmetry at large distances, while the latter case corresponds to solutions with unbroken symmetry at large distances, and for which there is therefore no topological stability argument.

The energy of the self-dual solutions is given by the saturated lower bound (4.17) which may be combined with the gauge field solution (4.23) to give

$$E_{\text{SD}} = \frac{v^2}{\kappa} \int d^2x\, B$$

$$= \frac{v^2}{\kappa} \int d^2x \left(\epsilon^{ij} \partial_i \partial_j \omega - \nabla^2 ln|\phi|^2 \right) \qquad (4.27)$$

To evaluate this self-dual energy we must distinguish between the two types of solutions in (4.26). For the topological solutions (4.26a), $ln(|\phi|^2/v^2)$ decreases exponentially at spatial infinity since

$$\delta \equiv 1 - \frac{|\phi|^2}{v^2} \qquad (4.28)$$

satisfies a massive equation

$$(-\nabla^2 + m^2)\delta = 0 \qquad (4.29)$$

Therefore, the self-dual energy of such a topological solution is determined by the discontinuity of the phase ω of the matter field in (4.22)

$$E = \frac{v^2}{\kappa} \left[\omega(\theta = 2\pi) \big|_{r=\infty} - \omega(\theta = 0) \big|_{r=\infty} \right] \qquad (4.30)$$

This energy is nonzero only if ω is discontinuous, and is quantized if we insist that the matter field ϕ is single-valued. Then the flux, charge and energy can all be expressed in terms of this integer discontinuity:

$$\Phi = 2\pi n \qquad (4.31a)$$

$$Q^0 = 2\pi\kappa n \qquad (4.31b)$$

$$E = \frac{v^2}{\kappa} 2\pi n \qquad (4.31c)$$

Note that the discontinuities in ω correspond to the singularities in $\vec{\nabla} \times \vec{\nabla}\omega$ in the finite spatial plane, and that these discontinuities cancel the singularities in the B field which correspond in turn to the singularities of $\nabla^2 ln|\phi|$. Such singularities arise from the zeros of $|\phi|$ in the finite plane, and are necessary for topological stability because without them it would be possible to deform ϕ continuously to the uniform symmetry-breaking configuration $|\phi|^2 = v^2$. Physically, these zeros correspond to the *locations* of the vortices, and at the self-dual point (with potential (4.16)) the vortices do not interact since the energy is the sum of individual contributions from each vortex at each zero of $|\phi|$.

We may be more explicit if we consider the special case of radially symmetric solutions. By a gauge transformation we can write the matter field as

$$\phi(\vec{x}) = f(r)e^{in\theta} \tag{4.32}$$

where n must be an integer for ϕ to be single-valued. Then from (4.23) the self-dual vector potential is

$$A_i = \epsilon^{ij}\frac{\hat{r}^j}{r}\left(\frac{rf'}{f} - n\right) \tag{4.33}$$

As $r \to \infty$ we are left with $A_\theta \sim n/r$ which leads to the quantized flux as in (4.31a). At the origin $f(r)$ must have an nth order zero

$$f(r) \sim r^n \tag{4.34}$$

in order for $\vec{A}$ to be regular there. The detailed asymptotic behavior of such radial solutions has been investigated numerically in [129] and some existence theorems have been proven in [240]. In particular, in [240] Wang has used complex analytic and variational techniques similar to those used in the study of the Nielsen-Olesen vortices of the abelian Higgs model [137] to prove that the abelian self-dual Chern-Simons equations (4.19) possess topological multi-vortex solutions which may be characterized by a finite set of zeros of the scalar field ϕ. In the neighborhood of each such zero z_k ($k = 1 \ldots m$), the scalar field ϕ can be written as

92

$$\phi(z) = (z - z_k)^{n_k} h_k(z) \tag{4.35}$$

where $h_k(z)$ is a smooth function with $h_k(z_k) \neq 0$. The total vortex number N is simply the sum of the orders of the zeros

$$N = \sum_{k=1}^{m} n_k, \tag{4.36}$$

and the zeros may be identified with the locations of the vortices. In addition, Spruck and Yang [226] have found an efficient constructive and numerical method for generating such multi-vortex solutions, permitting a more detailed and explicit study of their properties.

The angular momentum of a self-dual solution is

$$
\begin{aligned}
M &= \int d^2x \, \vec{x} \times \vec{\mathcal{P}} \\
&= \frac{1}{2} \int d^2x \, \vec{x} \times \left[(D_0\phi)^* \vec{D}\phi + \left(\vec{D}\phi\right)^* D_0\phi \right] \\
&= \frac{i}{2\kappa} \int d^2x \, \left(|\phi|^2 - v^2\right) \vec{x} \times \left(\phi^* \vec{D}\phi - (\vec{D}\phi)^*\phi\right) \\
&= \frac{\kappa}{2} \int d^2x \, B \left[(\vec{x} \times \vec{A} - \vec{x} \times \vec{\nabla} arg\phi\right]
\end{aligned}
\tag{4.37}
$$

Away from the zeros of $|\phi|$ we may replace B by $-\nabla^2 ln|\phi|$, so that for a radially symmetric self-dual solution

$$M = \pi\kappa \int_0^\infty dr \, \frac{d}{dr}\left(r\frac{d}{dr} ln\left(\frac{f}{v}\right)\right)^2 \tag{4.38}$$

The only contribution comes from the origin, where f behaves as in (4.34), leading to

$$M = -\pi\kappa n^2 = -\frac{1}{4\pi\kappa}(\mathcal{Q}^0)^2. \tag{4.39}$$

Nontopological solutions to the relativistic self-duality equations (4.19) may be obtained by considering solutions to (4.24) such that $|\phi| \to 0$ at spatial infinity. Then, at infinity

93

$$\nabla^2 ln|\phi|^2 \sim -\frac{4v^2}{\kappa^2}|\phi|^2 \qquad (4.40)$$

so that

$$\phi_n \sim r^{-\epsilon_n} \qquad (4.41)$$

where the exponent $\epsilon_n > 2$ and n refers to the number of zeros of ϕ_n in the finite spatial plane. This power-law fall-off leads to additional contributions to the various global quantities, beyond the contributions coming from the vorticity of ϕ. Indeed, for these nontopological solutions

$$\Phi = 2\pi(n + \epsilon_n) \qquad (4.42a)$$

$$Q^0 = 2\pi\kappa(n + \epsilon_n) \qquad (4.42b)$$

$$E = \frac{v^2}{\kappa}2\pi(n + \epsilon_n) \qquad (4.42c)$$

For radially symmetric solutions, the angular momentum is again given by (4.38), but now there is an additional contribution from infinity

$$M = -\pi\kappa n^2 + \pi\kappa\epsilon_n^2$$

$$= -\frac{(Q^0)^2}{4\pi\kappa} + \epsilon_n Q^0 \qquad (4.43)$$

A zero-mode analysis [129] of the spectrum of small fluctuations (analogous to Weinberg's analysis of the Landau-Ginzburg system [245]) indicates that the index

$$\mathcal{I}(D) \equiv dim ker D_+ D_- - dim ker D_- D_+$$

94

$$= \begin{cases} 2n, & |\phi|^2 \to v^2 \text{ as } r \to \infty \\ \\ 2n + 2[\epsilon_n], & |\phi|^2 \to 0 \text{ as } r \to \infty \end{cases} \tag{4.44}$$

where $[\epsilon_n]$ denotes the integer part of ϵ_n. This index counts the number of fluctuation degrees of freedom, and in the topological case the number $2n$ is identified with the number of parameters required to describe the locations of the n vortices. In the nontopological case the counting is less clear, as are the stability properties [128,129,16,146,225].

D. Nonrelativistic Limit

To consider the nonrelativistic limit of the relativistic self-dual Chern-Simons theory with Lagrange density (4.1) we must first reintroduce the speed of light c, which had previously been set to unity. The dimensionally corrected Lagrange density has matter contribution

$$\mathcal{L}_{\text{matter}} = -(D_\mu \phi)^\dagger D^\mu \phi - \frac{1}{c^4 \kappa^2} |\phi|^2 \left(|\phi|^2 - v^2\right)^2 \tag{4.45}$$

where $D_\mu \phi = \partial_\mu + \frac{i}{c} A_\mu \phi$. The scalar field mass term in the corresponding potential is

$$\frac{v^4}{c^4 \kappa^2} |\phi|^2 \equiv m^2 c^2 |\phi|^2 \tag{4.46}$$

so that the scalar field excitation has mass

$$m = \frac{v^2}{c^3 |\kappa|} \tag{4.47}$$

In order to maintain a finite mass for the scalar field in the nonrelativistic limit, the $c \to \infty$ limit must be accompanied by the limit $v^2 \to \infty$ in such a way that

95

$$\frac{v^2}{c^3} = \text{constant} \tag{4.48}$$

This is another reason why it was necessary to include a mass term in the self-dual potential (4.16). Without such a mass term, and the associated mass scale v^2, it would not be possible to define a sensible nonrelativistic limit.

The nonrelativistic limit now proceeds in the standard manner [128]. Separating the rest-mass energy as

$$\phi = \frac{1}{\sqrt{2m}} e^{-imc^2 t} \psi \tag{4.49}$$

and writing $x^0 = ct$, the matter Lagrange density (keeping dominant terms in powers of $1/c$) becomes

$$\mathcal{L}_{\text{matter}}^{\text{NR}} = i\psi^* \left(\partial_t + \frac{i}{c} A_0 \right) \psi - \frac{1}{2m} \left| \left(\vec{\nabla} + \frac{i}{c} \vec{A} \right) \psi \right|^2 + \frac{1}{2mc|\kappa|} |\psi|^4 \tag{4.50}$$

Note that the $|\psi|^2$ term in the potential is cancelled by a term coming from $|D_0 \phi|^2$ (as is usual in a nonrelativistic limit), and the sixth-order term in the potential becomes

$$-\frac{1}{8m^3 c^4 \kappa^2} |\psi|^6 \tag{4.51}$$

which may be dropped relative to other terms in the nonrelativistic $c \to \infty$ limit.

We observe that the nonrelativistic limit matter Lagrange density (4.50) coincides precisely with the self-dual form of the matter part of the nonrelativistic Chern-Simons Lagrange density (2.33), with the form and the strength of the attractive $|\psi|^4$ potential (2.86) being determined by the nonrelativistic limit.

One may also consider the nonrelativistic limit at the level of the equations of motion and the self-duality equations. With factors of

96

c restored, the relativistic self-dual Chern-Simons equations (4.19) read:

$$D_- \phi = 0 \qquad (4.52a)$$

$$F_{12} = -\frac{2}{c^3 \kappa^2}|\phi|^4 + \frac{2v^2}{\kappa^2 c^3}|\phi|^2 \qquad (4.52b)$$

Writing the fields as in (4.49) and taking the nonrelativistic limit $c \to \infty$ with m fixed these become

$$D_- \psi = 0 \qquad (4.53a)$$

$$
\begin{aligned}
F_{12} &= -\frac{1}{2m^2\kappa^2 c^3}|\psi|^4 + \frac{v^2}{\kappa^2 c^3 m}|\psi|^2 \\
&= -\frac{1}{2m^2\kappa^2 c^3}|\psi|^4 + \frac{1}{\kappa}|\psi|^2 \\
&\to \frac{1}{\kappa}|\psi|^2 \qquad (4.53b)
\end{aligned}
$$

The quartic term in (4.53b) is suppressed by a factor of $1/c^3$ and so is dropped. Further, the first-order self-dual condition (4.18b) becomes, in terms of the nonrelativistic field ψ :

$$i\left(\partial_t + \frac{i}{c}A_0\right)\psi = -\frac{1}{2mc\kappa}|\psi|^2\psi \qquad (4.54)$$

This is just the gauged nonlinear Schrödinger equation (2.4a) with the critical coupling (2.32) for the nonrelativistic potential in (2.33), and with ψ satisfying the nonrelativistic self-duality equation (2.34a). Identifying A_0 as in (2.31) yields

$$\partial_t \psi = 0 \qquad (4.55)$$

which are the *static* self-dual solutions of the nonrelativistic self-dual Chern-Simons system, as discussed in Section II C.

97

E. Symmetry Breaking and the Chern–Simons Higgs

Mechanism

Regarded as a symmetry breaking problem, the relativistic self-dual Chern-Simons system with Lagrange density (4.1) is rather different from a conventional Higgs system. First, in $3+1$ dimensional field theory one most commonly considers symmetry breaking potentials of ϕ^4 form, but here in $2+1$ dimensions we consider a (renormalizable) sixth order potential (4.16). In the unbroken phase, about the trivial minimum at $|\phi| = 0$, the complex scalar field ϕ has two real massive degrees of freedom, each of mass

$$m = \frac{v^2}{\kappa} \tag{4.56}$$

In the broken phase, about the nontrivial minimum at $|\phi| = |v|$, we expand ϕ as

$$\phi = \phi' + |v|$$

$$= \frac{1}{\sqrt{2}} \left(\phi'_1 + i\phi'_2 \right) + |v| \tag{4.57}$$

Then the quadratic part of $V(\phi' + |v|)$ is

$$V_{\text{quad}}(\phi' + |v|) = \frac{1}{2} \left(\frac{2v^2}{\kappa} \right)^2 \phi_1^2 \tag{4.58}$$

so there is just one real massive scalar field of mass

$$\mu = \frac{2v^2}{\kappa}$$

$$= 2m \tag{4.59}$$

Notice that the mass of the scalar ("Higgs") field in the broken phase is *twice* the mass of the scalar fields in the unbroken phase. This is a consequence of the particular form of the self-dual potential (4.16).

The second, and more significant, difference between the self-dual Chern-Simons system (4.1) and a conventional Higgs system is that the Higgs mechanism for generating massive gauge degrees of freedom behaves very differently in a 2 + 1 dimensional theory with a Chern- Simons term present for the gauge field. In 2 + 1 dimensional gauge theories there are in fact four separate possibilities for producing massive gauge field excitations:

- The gauge masses are produced by the Higgs mechanism alone.

- The gauge masses are produced by both a Maxwell and a Chern- Simons term, with no symmetry breaking.

- The gauge masses are produced by both a Maxwell and a Chern- Simons term, plus symmetry breaking.

- The gauge masses are produced by a Chern-Simons term alone, plus symmetry breaking.

The first case is standard. To study the second case, consider a Maxwell-Chern-Simons theory (with no matter fields, and consequently no symmetry breaking) with Lagrange density

$$\mathcal{L}_{\text{MCS}} = -\frac{1}{4e^2} F_{\mu\nu} F^{\mu\nu} + \frac{\mu}{2e^2} \epsilon^{\mu\nu\rho} A_\mu \partial_\nu A_\rho \qquad (4.60)$$

Note that the electric charge e has been reintroduced in order to balance dimensions, and e^2 has dimensions of mass in 2 + 1-dimensional spacetime. The Chern-Simons coupling

$$\kappa = \frac{\mu}{e^2} \qquad (4.61)$$

is dimensionless, with μ also having dimensions of mass. The Lagrange density (4.60) is quadratic, even when we include a covariant gauge-fixing term

99

$$\mathcal{L}_{\text{g.f.}} = -\frac{1}{2\xi e^2} \left(\partial_\mu A^\mu \right)^2 \tag{4.62}$$

The full quadratic Lagrange density for the gauge fields is

$$\mathcal{L}_{\text{MCS}} = \frac{1}{2e^2} A_\mu \left\{ [\nabla^2 g^{\mu\nu} - \left(1 - \frac{1}{\xi} \right) \partial^\mu \partial^\nu] - \mu \epsilon^{\mu\nu\rho} \partial_\rho \right\} A_\nu$$
$$\equiv -\frac{1}{2} A_\mu \left(\Delta^{\mu\nu} \right)^{-1} A_\nu \tag{4.63}$$

Then the momentum space gauge propagator is

$$\Delta_{\mu\nu} = e^2 \left(\frac{p^2 g_{\mu\nu} - p_\mu p_\nu - i\mu \epsilon_{\mu\nu\rho} p^\rho}{p^2 (p^2 - \mu^2)} + \xi \frac{p_\mu p_\nu}{(p^2)^2} \right) \tag{4.64}$$

From the poles of this propagator we identify one physical excitation mode of mass μ, and (as expected) an unphysical massless mode which decouples in a physical gauge [46,219].

Now consider the effect of coupling this Maxwell-Chern-Simons Lagrange density (4.60) to a complex scalar field ϕ with a symmetry breaking potential $V(|\phi|)$

$$\mathcal{L}_{\text{MCSH}} = -\frac{1}{4e^2} F_{\mu\nu} F^{\mu\nu} + \frac{\mu}{2e^2} \epsilon^{\mu\nu\rho} A_\mu \partial_\nu A_\rho - (D_\mu \phi)^* D^\mu \phi - V(|\phi|) \tag{4.65}$$

where $V(|\phi|)$ has some nontrivial vacuum $<\phi>_{(0)}$. In the broken vacuum we consider fluctuations of the scalar field ϕ about its vacuum expectation value $|v|$, and so the scalar kinetic term $|D_\mu \phi|^2$ leads to an additional quadratic term in the gauge field

$$v^2 A_\mu A^\mu \tag{4.66}$$

Thus, the quadratic part of the gauge Lagrange density in the broken vacuum is

$$\mathcal{L}_{quad} = -\frac{1}{4e^2} F_{\mu\nu} F^{\mu\nu} + \frac{\mu}{2e^2} \epsilon^{\mu\nu\rho} A_\mu \partial_\nu A_\rho - v^2 A_\mu A^\mu \tag{4.67}$$

The momentum space propagator, with a covariant 't Hooft gauge fixing term is now [205,110]

$$\Delta_{\mu\nu} =$$

$$\frac{e^2(p^2 - m_H^2)}{(p^2 - m_+^2)(p^2 - m_-^2)}\left[g_{\mu\nu} - \frac{p_\mu p_\nu}{(p^2 - \xi m_H^2)} - i\frac{\mu\epsilon_{\mu\nu\rho}p^\rho}{(p^2 - m_H^2)}\right]$$

$$+ e^2\xi p_\mu p_\nu \frac{(p^2 - \mu^2 - m_H^2)}{(p^2 - m_+^2)(p^2 - m_-^2)(p^2 - \xi m_H^2)} \tag{4.68}$$

where

$$m_H^2 = 2e^2v^2 \tag{4.69}$$

is the usual Higgs mass scale (squared) and

$$D(p^2) = (p^2 - m_H^2)^2 - \mu^2 p^2$$

$$\equiv (p^2 - m_+^2)(p^2 - m_-^2) \tag{4.70}$$

where

$$m_\pm^2 = m_H^2 + \frac{\mu^2}{2} \pm \frac{\mu}{2}\sqrt{\mu^2 + 4m_H^2} \tag{4.71}$$

or

$$m_\pm = \frac{\mu}{2}\left(\sqrt{1 + \frac{4m_H^2}{\mu^2}} \pm 1\right) \tag{4.72}$$

From this propagator we identify *two* physical mass poles at

$$p^2 = m_\pm^2 \tag{4.73}$$

as well as the (expected) unphysical gauge-variant mass pole at $p^2 = \xi m_H^2$ [205]. The appearance of these two mass poles may also

101

be understood in terms of a self-dual factorization of the Maxwell-Chern-Simons Proca equations [200], analogous to the dual formulation of the pure Maxwell-Chern-Simons theory [47].

The counting of degrees of freedom goes as follows. In the unbroken vacuum, the complex scalar field has two real massive degrees of freedom and the gauge field has one massive excitation (with mass coming from the Chern-Simons term). In the broken vacuum, one component of the scalar field (the "Goldstone boson") combines with the longitudinal part of the gauge field to produce a new massive gauge degree of freedom. Thus, in the broken vacuum there are *two* massive gauge degrees of freedom and *one* real massive scalar degree of freedom (the "Higgs boson").

The self-dual Chern-Simons theory with Lagrange density (4.1) has just a Chern-Simons term (and no Maxwell term) for the gauge field in the Lagrange density. The limit in which the Maxwell term is decoupled may be achieved by taking

$$e^2 \to \infty \qquad\qquad \mu \to \infty \qquad\qquad \kappa \equiv \frac{\mu}{e^2} = \text{fixed} \qquad (4.74)$$

with the dimensionless Chern-Simons coupling parameter κ (4.61) kept fixed. Such a limit was introduced in [104]. This leads to a Lagrange density of the self-dual Chern-Simons form

$$\mathcal{L}_{\text{CSH}} = \frac{\mu}{2e^2} \epsilon^{\mu\nu\rho} A_\mu \partial_\nu A_\rho - (D_\mu \phi)^\dagger D^\mu \phi - V(|\phi|) \qquad (4.75)$$

Then the propagator (4.68) reduces in this limit to

$$\Delta_{\mu\nu} = \frac{1}{p^2 - (2v^2/\kappa)^2} \left(\frac{2v^2}{\kappa^2} g_{\mu\nu} - \frac{1}{2v^2} p_\mu p_\nu + \frac{i}{\kappa} \epsilon_{\mu\nu\rho} p^\rho \right) \qquad (4.76)$$

From this propagator we deduce a *single* massive pole at

$$p^2 = \left(\frac{2v^2}{\kappa} \right)^2 \qquad (4.77)$$

The counting of degrees of freedom is different in this Chern-Simons-Higgs model. In the unbroken vacuum the gauge field is nonpropagating, and so there are just the two real scalar modes of the scalar field ϕ. In the broken vacuum, one component of the scalar field (the "Goldstone boson") combines with the longitudinal part of the gauge field to produce a massive gauge degree of freedom. Thus, in the broken vacuum there is *one* massive gauge degree of freedom and *one* real massive scalar degree of freedom (the "Higgs boson"). This may also be deduced from the mass formulae (4.72) which, in the limit (4.74) tend to

$$m_- \to \frac{2v^2}{\kappa} \qquad\qquad m_+ \to \infty \qquad (4.78)$$

so that one mass decouples to infinity. This pure Chern-Simons-Higgs mechanism was first considered by Deser and Yang [48], and by Wen and Zee [246].

A simple physical picture of the masses (4.72) generated by the Chern-Simons Higgs mechanism may be obtained by a Schrödinger representation analysis of the quadratic Lagrange density (4.67). In the Schrödinger representation one constructs the corresponding functional Hamiltonian in the physical subspace, and so this approach is particularly suited to the identification of physical excitations [123]. Generically, the diagonalized quadratic Hamiltonian has the form

$$\mathcal{H}_{\text{quad}} = \sum_\alpha a_\alpha^\dagger \left[\sqrt{-\nabla^2 + m_\alpha^2} \right] a_\alpha \qquad (4.79)$$

where the sum is over the physical modes of mass m_α. This diagonalization has been performed in [63] for both the Maxwell-Chern-Simons-Higgs and Chern-Simons-Higgs systems. However, if we are only interested in the *masses* of the excitations, then it is sufficient to make a zeroth-order spatial derivative expansion (*i.e.* neglecting all spatial derivatives), in which case the functional Schrödinger

103

representation reduces to the familiar Schrödinger representation of quantum mechanics. Physical *masses* of the field theory then appear as physical *frequencies* of the corresponding quantum mechanical system. In this limit, the quadratic Lagrange density (4.67) becomes

$$L = \frac{1}{2e^2}\dot{A}_i^2 + \frac{\mu}{2e^2}\epsilon^{ij}\dot{A}_i A_j + v^2 A_i A_i \qquad (4.80)$$

where now $A_i = A_i(t)$ is a function of time only. This model which has been studied in relation to 'Chern-Simons quantum mechanics' [56,213,60,44,84]. This Lagrangian is the quantum mechanical Lagrangian for a charged particle of mass $\frac{1}{e^2}$ moving in a uniform magnetic field of strength $\frac{\mu}{e^2}$ and a harmonic potential well of frequency $\sqrt{2}ev$. Such a quantum mechanical model is exactly solvable, and is well-known to separate into two separate harmonic oscillator systems of characteristic frequencies

$$\omega_{\pm} = \frac{\omega_c}{2}\left(\sqrt{1 + \frac{4\omega^2}{\omega_c^2}} \pm 1\right) \qquad (4.81)$$

where ω_c is the cyclotron frequency corresponding to the magnetic field and ω is the harmonic well frequency. Comparing with the Maxwell-Chern-Simons-Higgs Lagrangian (4.80) we see that the characteristic frequencies $\omega_{\pm}$ do indeed agree with the mass poles $m_{\pm}$ in (4.72), which were identified from the covariant gauge propagator. The pure Chern-Simons Higgs limit (4.74) corresponds to the physical limit in which the magnetic field strength dominates so that

$$\omega_- \to \frac{\omega^2}{\omega_c} = \frac{2v^2}{\kappa} = m_- \qquad\qquad \omega_+ \to \infty \qquad (4.82)$$

Thus, this limit corresponds to a truncation of the physical Hilbert space in which the dynamics is projected onto the lowest Landau level. Such a limit is physically relevant, and of great interest, in

104

applications to theories of the quantum Hall effect [98,257,229,60]. It has also been studied recently in the context of cold trapped Rydberg atoms [15], another physical setting which has the potential of realizing the Chern-Simons limit (4.74).

F. Self-Dual Chern–Simons Theories and Extended Supersymmetry

It is a well-documented fact that there is a deep connection between self-dual models and extended supersymmetry [248,111,112]. This can be seen both in terms of the structure of the Lagrange density and in terms of the relation of the topological Bogomol'nyi bound to a topological quantum number which appears as the central charge in the extended supersymmetry algebra. The relativistic self-dual Chern-Simons models provide another explicit example of this general phenomenon.

We begin by illustrating the role of extended supersymmetry in the abelian relativistic self-dual Chern-Simons systems. Additional subtleties arise for nonabelian theories, but we defer comments about this until the end of this Section. The basic result [165] we wish to convey is that the requirement of extended supersymmetry essentially fixes the form of the superpotential in such a way that the bosonic part of the Lagrange density takes the form (4.1) with the potential having its self-dual form (4.16).

Consider an $N = 1$ supersymmetric generalization of the Chern-Simons Higgs model (4.1). The $N = 1$ superfield formalism for $2+1$ dimensional spacetime is discussed in [94], and other interesting features of $2+1$ dimensional supersymmetry can be found in [224,1]. A complete discussion of supersymmetry in $2+1$ dimensions is beyond the scope of these Lecture Notes, so I refer the interested reader to

105

[224,94,1] for further details. Here we shall content ourselves with the illustration of the significance of extended supersymmetry in the self-dual Chern-Simons systems. We follow closely the analysis of [165]. The matter fields are described by a complex scalar superfield Φ, which contains a complex scalar field ϕ, a complex spinor field ψ, and an auxiliary complex scalar field F. There is also a real gauge-spinor superfield Γ_α which contains a real gauge field A_μ and a real (Majorana) spinor field λ_α.

$$\Phi < -- > (\phi, \psi, F)$$

$$\Gamma_\alpha < -- > (A_\mu, \lambda_\alpha) \tag{4.83}$$

Explicitly, the complex scalar superfield is expanded as

$$\Phi(x, \theta) = \phi(x) + \theta^\alpha \psi_\alpha(x) - \theta^2 F(x) \tag{4.84}$$

where all the spinor indices α are raised and lowered with respect to the second rank antisymmetric symbol

$$C_{\alpha\beta} = -C_{\beta\alpha} = -C^{\alpha\beta} = \begin{pmatrix} 0 & -i \\ i & 0 \end{pmatrix}$$

$$\psi_\alpha = \psi^\beta C_{\beta\alpha}$$

$$\psi^\alpha = C^{\alpha\beta} \psi_\beta$$

$$\theta^2 = \frac{1}{2} \theta^\alpha \theta_\alpha \tag{4.85}$$

The superderivative is defined as [94]

$$D_\alpha = \frac{\partial}{\partial\theta^\alpha} + i\theta^\beta (\tilde\sigma_\mu)_{\beta\alpha} \, \partial^\mu \tag{4.86}$$

where $\tilde\sigma_\mu \equiv (-1, \sigma_3, \sigma_1)$ and we take the Dirac γ matrices to be

$$\gamma_\mu = \sigma_2 \bar{\sigma}_\mu \qquad (4.87)$$

where $\bar{\sigma}_\mu \equiv (1, \sigma_3, \sigma_1)$. These Dirac matrices satisfy $\{\gamma_\mu, \gamma_\nu\} = -2\eta_{\mu\nu}\mathbf{1}$.

Minimal coupling with the gauge superfield Γ is achieved through the gauge supercovariant derivative

$$\nabla_\alpha = D_\alpha + i\Gamma_\alpha \qquad (4.88)$$

The component fields making up the complex scalar superfield Φ may be expressed as

$$\phi = \Phi| \qquad \psi_\alpha = \nabla_\alpha \Phi| \qquad F = \nabla^2 \Phi| \qquad (4.89)$$

where the notation $(\ldots)|$ means $(\ldots)$ evaluated with the supercoordinates θ set to zero [94]. The components of the real superfield Γ_α are expressed as

$$A_\mu = -\frac{i}{2} D_\alpha (\bar{\sigma}_\mu)^{\alpha\beta} \Gamma_\beta|$$

$$\lambda_\alpha = \frac{1}{2} D^\beta D_\alpha \Gamma_\beta| \qquad (4.90)$$

The corresponding field-strength superfield is

$$W_\alpha = \frac{1}{2} D^\beta D_\alpha \Gamma_\beta \qquad (4.91)$$

which satisfies $D^\alpha W_\alpha = 0$.

We can now write an $N = 1$ superspace action

$$S = \int d^3x d^2\theta \left[-\frac{\kappa}{4} \Gamma^\alpha W_\alpha - \frac{1}{2} (\nabla^\alpha \Phi)^* \nabla_\alpha \Phi + f(\Phi^*\Phi) \right] \qquad (4.92)$$

where we keep the superpotential $f(\Phi^*\Phi)$ unspecified at this point. The first term,

$$-\frac{\kappa}{4} \Gamma^\alpha W_\alpha \qquad (4.93)$$

is the supersymmetric version of the Chern-Simons term, as first discussed by Siegel [224]. After a straightforward (but tedious) expansion one can rewrite this action in terms of component fields as

$$S = \int d^3x \left[\frac{\kappa}{2} \epsilon^{\mu\nu\rho} A_\mu \partial_\nu A_\rho - |D_\mu \phi|^2 - i\psi^{*\alpha} (\gamma^\mu)_\alpha^\beta D_\mu \psi_\beta \right.$$

$$-i \left(\psi^{*\alpha} \lambda_\alpha \phi - \lambda^\alpha \psi_\alpha \phi^* \right) + f'(|\phi|^2) \left(F^* \phi + F \phi^* \right)$$

$$+ \frac{1}{2} f''(|\phi|^2) \left(\phi^2 \psi^{*\alpha} \psi_\alpha^* - \phi^{*2} \psi^\alpha \psi_\alpha \right) - \frac{\kappa}{2} \lambda^\alpha \lambda_\alpha$$

$$\left. + \psi^{*\alpha} \psi_\alpha \left(f'(|\phi|^2) + |\phi|^2 f''(|\phi|^2) \right) + F^* F \right] \qquad (4.94)$$

where $D_\mu = \partial_\mu + iA_\mu$ is the usual covariant derivative. We recognize the first term in this action as the Chern-Simons term for the gauge field. This action is invariant under the supersymmetry transformations

$$\delta\phi = -\eta^\alpha \psi_\alpha$$

$$\delta A_\mu = -i\eta^\alpha (\gamma_\mu)_\alpha^\beta \lambda_\beta$$

$$\delta\lambda_\alpha = \frac{i}{2} \epsilon^{\mu\nu\rho} F_{\nu\rho} (\gamma_\mu)_\alpha^\beta \eta_\beta$$

$$\delta\psi_\alpha = -i(\gamma_\mu)_\alpha^\beta \eta_\beta D_\mu \phi + \eta_\alpha F$$

$$\delta F = -i\eta^\alpha (\gamma_\mu)_\alpha^\beta D_\mu \psi_\beta + i\eta^\alpha \lambda_\alpha \phi \qquad (4.95)$$

where η is a real infinitesimal spinor parameter. The fields λ_α, F and F^* are auxiliary fields which may be eliminated via their 'equations of motion'

$$\lambda_\alpha = -\frac{i}{\kappa} \left(\psi_\alpha^* \phi - \psi_\alpha \phi^* \right)$$

108

$$F = -\phi f'(|\phi|^2)$$

$$F^* = -\phi^* f'(|\phi|^2) \tag{4.96}$$

The resulting component-field action is

$$S = \int d^3x \left[\frac{\kappa}{2} \epsilon^{\mu\nu\rho} A_\mu \partial_\nu A_\rho - |D_\mu \phi|^2 - i\psi^{*\alpha} (\gamma^\mu)_\alpha^\beta D_\mu \psi_\beta \right.$$

$$+ \frac{1}{2} \left(f''(|\phi|^2) + \frac{1}{\kappa} \right) \left[\phi^2 \psi^{*\alpha} \psi_\alpha^* - \phi^{*2} \psi^\alpha \psi_\alpha \right]$$

$$+ \psi^{*\alpha} \psi_\alpha \left\{ f'(|\phi|^2) + |\phi|^2 \left(f''(|\phi|^2) - \frac{1}{\kappa} \right) \right\}$$

$$\left. - |\phi|^2 \left(f'(|\phi|^2) \right)^2 \right] \tag{4.97}$$

This action is invariant under the $N = 1$ supersymmetry transformation

$$\delta A_\mu = -\frac{1}{\kappa} \left(\psi^{*\alpha} (\gamma_\mu)_\alpha^\beta \eta_\beta \phi + \eta^\alpha (\gamma_\mu)_\alpha^\beta \psi_\beta \phi^* \right)$$

$$\delta \phi = -\eta^\alpha \psi_\alpha$$

$$\delta \psi_\alpha = -i (\gamma_\mu)_\alpha^\beta \eta_\beta D_\mu \phi - \eta_\alpha \phi f'(|\phi|^2) \tag{4.98}$$

An $N = 2$ supersymmetric generalization of this action must conserve fermion number, so the fermion number violating term in (4.97) must be cancelled by a suitable choice of the superpotential function f. This forces

$$f'' = -\frac{1}{\kappa} \tag{4.99}$$

which means that the superpotential is determined to be (up to an unimportant additive constant) of the form

$$f(\Phi^* \Phi) = -\frac{1}{2\kappa} \left((\Phi^* \Phi) - v^2 \right)^2 \tag{4.100}$$

109

Thus, the fermion number conserving form of the action (4.97) is

$$S = \int d^3x \left[\frac{\kappa}{2} \epsilon^{\mu\nu\rho} A_\mu \partial_\nu A_\rho - |D_\mu \phi|^2 + i\bar{\psi}\gamma^\mu D_\mu \psi \right.$$

$$\left. - \frac{1}{\kappa^2} |\phi|^2 \left(|\phi|^2 - v^2 \right)^2 - \frac{1}{\kappa} \left(3|\phi|^2 - v^2 \right) \bar{\psi}\psi \right] \qquad (4.101)$$

where $i\psi = \begin{pmatrix} \psi_1 \\ \psi_2 \end{pmatrix}$ and $\bar{\psi} = \psi^\dagger \gamma^0$.

We observe that the bosonic portion of this action (4.101) is identical to the action of the self-dual Chern-Simons Higgs system (4.1) with self-dual potential (4.16). Furthermore, we note that in the broken vacuum in which the scalar field ϕ has a vacuum expectation value $< \phi^2 > = v^2$, the fermion field has a mass

$$m_{\text{fermion}} = \frac{2v^2}{\kappa} \qquad (4.102)$$

which is degenerate with both the gauge and (Higgs) scalar mass (4.59) in the broken vacuum. The degeneracy of these masses is a reflection of the supersymmetric multiplet structure of the theory.

Thus, in this abelian theory, the requirement of $N = 2$ supersymmetry (which in the above is obtained by cancelling the fermion number violating piece) uniquely determines the superpotential, and when reduced to its component-field form we note that the resulting bosonic potential has the self-dual form. However, some qualifying comments are in order, as in this regard the abelian theory is rather special. In more general self-dual Chern-Simons theories (for example, with many $U(1)$ matter fields, or with a local nonabelian gauge symmetry) the relationship between the superpotential and the requirement of $N = 2$ supersymmetry is richer. This can be seen most clearly by constructing a manifestly $N = 2$ supersymmetric superspace formulation of the self-dual Chern-Simons system [120,95,197]. Such a formulation is arguably the most natural with which to analyze the extended supersymmetry. In this case certain special features of the supermultiplet structure of $2+1$ dimensional models come into play. In particular, in an $N=2$ superfield Lagrangian, a scalar supermultiplet is represented by a chiral superfield; but it is not possible to write down an explicitly $U(1)$ invariant *potential* term for *one* $N=2$ chiral superfield [94]. Therefore, when reduced to the component fields (or in an $N=1$ superfield description [228]) the superpotentials can only have come form the minimal coupling of the $N=2$ matter superfield to an $N=2$ gauge field. This greatly restricts the possible form of the $N=2$ superfield Lagrangian. This is exactly what was done in the above discussion (when translated into a manifest $N=2$ formulation [120]) for the abelian theory, and which led to a unique fixing of the superpotential to its self-dual form (4.100) as a result of the requirement of $N=2$ supersymmetry. However, when one includes multiple $U(1)$ matter fields, or local nonabelian gauge symmetries, there *do* exist explicitly invariant $N=2$ superpotential terms, so that $N=2$ supersymmetry now permits new couplings not considered in the above abelian discussion. This would seem to suggest that the requirement of $N=2$ supersymmetry is not sufficient to prescribe the self-dual form of the interactions in such models. Nevertheless, we shall see that the nonabelian self-dual Chern-Simons theories (which will be introduced in the next Chapter) possess a global $U(1)$ symmetry together with their local nonabelian gauge symmetry, and this is enough to restore the correspondence between self-duality and the requirement of $N=2$ supersymmetry, as in the abelian models. Indeed, this relationship between self-duality and extended supersymmetry has been taken a step further by Kao and K. Lee who have constructed maximally supersymmetric $N=3$ models corresponding to self-dual Chern-Simons-Higgs theories [141,142].

An important consequence of this connection between self-duality and extended supersymmetry is that the topological lower bound for the energy appears in the supersymmetry algebra satisfied by the supercharges. Once again, this is a particular example of a general feature of self-dual theories [111,112], and we illustrate it here for the self-dual Chern-Simons theories following the analysis of Lee, Lee and Weinberg [165]. The spinor supercharge Q which generates the supersymmetry transformation (4.98) is

$$Q = \int d^2x \left[\gamma^\mu \gamma^0 \psi D_\mu \phi^* + i\gamma^0 \psi \phi^* f'(|\phi|^2) \right] \qquad (4.103)$$

As a consequence of the canonical commutation relations between the fields, these supercharge operators satisfy a centrally-extended supersymmetry algebra:

$$\{Q_\alpha, \bar{Q}^\beta\} = (\gamma^\mu)_\alpha{}^\beta P_\mu - \delta_\alpha{}^\beta T \qquad (4.104)$$

where P_μ is the momentum generator and T is the central charge of the supersymmetry algebra

$$T = \int d^2x \left[F_{12}|\phi|^2 - \frac{i}{\kappa} \left(|\phi|^2 - v^2 \right) \left(\phi^* D_0\phi - \phi(D_0\phi)^* - i\bar{\psi}\gamma^0\psi \right) \right] \qquad (4.105)$$

Using the Gauss law constraint for the supersymmetric action (4.101),

$$\kappa F_{12} = i \left(\phi^* D_0\phi - \phi(D_0\phi)^* - i\bar{\psi}\gamma^0\psi \right), \qquad (4.106)$$

we find that the central charge is

$$T = v^2 \int d^2x\, F_{12} \qquad (4.107)$$

which is proportional to the magnetic flux. Then the supersymmetry algebra implies that

$$P^0 = T + \{Q_+, Q_+^\dagger\} \qquad\qquad (4.108)$$

where $Q_+ = \frac{1}{2}(1+\gamma^0)Q$. This means that the energy, P^0, has a lower bound given by T, and this lower bound agrees exactly with the lower bound (4.17) obtained previously by an explicit factorization of the energy density. Such a relation between the Bogomol'nyi bound and central charges in an associated supersymmetry algebra was first noted in other models [248], but has since been recognized as an important general property of self-dual theories [111,112]. For example, for an explicit detailed description of this feature in the self-dual Abelian Higgs model in $2 + 1$ dimensions, see [75].

G. Maxwell–Chern–Simons Model

Just as the nonrelativistic model of Chapter 2 was generalized in Section II H to include a Maxwell term for the gauge field, so the relativistic model described in this Chapter may be enlarged to incorporate a propagating topologically massive Maxwell-Chern-Simons gauge field [166]. In order to maintain a notion of self-duality, in which a lower bound for the energy is saturated by solutions to a set of self-duality equations, it is necessary to introduce an additional neutral scalar field. A related curved space model, which also exhibits self-duality, has been discussed in [29] in the context of Einstein gravity in $2 + 1$ dimensions (see also [238] for a curved space pure Chern-Simons model).

Consider the Lagrange density

$$\mathcal{L} = -\frac{1}{4e^2}F_{\mu\nu}F^{\mu\nu} + \frac{\mu}{2e^2}\epsilon^{\mu\nu\rho}A_\mu\partial_\nu A_\rho - |D_\mu\phi|^2$$

$$-\frac{1}{2e^2}(\partial_\mu N)^2 - V(\phi, N) \qquad\qquad (4.109)$$

where the potential $V(\phi, N)$ chosen to take the special self-dual form

$$V(\phi, N) = |\phi|^2 \left(N - \frac{e^2}{\mu}v^2\right)^2 + \frac{e^2}{2}\left(|\phi|^2 - \frac{\mu}{e^2}N\right)^2 \qquad (4.110)$$

Note that there are three separate mass scales: the Chern-Simons mass scale μ, the electric charge squared e^2, and the symmetry breaking scale v^2. As before, the Chern-Simons coupling parameter κ is the dimensionless ratio $\kappa = \mu/e^2$. The self-dual potential has two degenerate minima: a *symmetric* phase in which $< \phi >= 0$ and $< N >= 0$, and an *asymmetric* phase in which $< |\phi| >= v$ and $< N >= \frac{e^2 v^2}{\mu} = \frac{v^2}{\kappa}$.

In the symmetric phase the complex scalar field ϕ has mass

$$m_{\text{sc}} = \frac{e^2 v^2}{\mu} = \frac{v^2}{\kappa} \qquad (4.111)$$

while the neutral scalar field N and the gauge field have the common mass

$$m_N = m_{\text{gauge}} = \mu \qquad (4.112)$$

In the asymmetric phase the Chern-Simons-Higgs mechanism (discussed in Section IV E) produces two massive gauge degrees of freedom with masses $m_\pm$ given by

$$m_\pm^2 = 2e^2 v^2 + \frac{\mu^2}{2} \pm \frac{\mu}{2}\sqrt{\mu^2 + 8e^2 v^2} \qquad (4.113)$$

The quadratic part of the scalar field Lagrange density in the broken phase is

$$\mathcal{L}_{\text{quad}}^{\text{scalar}} = -\frac{1}{2e^2}(\partial_\mu N)^2 - \frac{1}{2}(\partial_\mu \phi_R)^2 - \frac{1}{2}(\partial_\mu \phi_I)^2 - e^2 v^2 \phi_R^2$$

$$+ \sqrt{2}\mu v N \phi_R - N^2\left(\frac{\mu^2}{2e^2} + v^2\right) \qquad (4.114)$$

where the fields have been shifted as

114

$$\phi \rightarrow \frac{1}{\sqrt{2}} (\phi_R + i\phi_I) + <\phi>$$

$$N \rightarrow N+ <N> \tag{4.115}$$

Diagonalizing, we see that the real fields N and ϕ_R combine into two massive scalar modes with masses $m_\pm$ equal to the gauge masses in (4.113). As is characteristic of self-dual models, the massive modes in each vacuum come in *pairs* (see also Section V D).

In the pure Chern-Simons limit, $e^2 \rightarrow \infty$, $\mu \rightarrow \infty$ (with $\kappa = \mu/e^2$ fixed), the Maxwell term drops out of the Lagrange density, and correspondingly the mass m_+ decouples to infinity. The scalar N field also decouples, as it is forced to be evaluated at

$$N = \frac{e^2}{\mu} |\phi|^2 \tag{4.116}$$

in which case the self-dual potential (4.110) becomes

$$V = \frac{1}{\kappa^2} |\phi|^2 \left(|\phi|^2 - v^2 \right)^2 \tag{4.117}$$

which is just the sixth-order self-dual potential (4.16) for the pure Chern-Simons system. Similarly, in the limit $\kappa \rightarrow 0$ in which the Chern-Simons term is removed, we see that N must be evaluated as

$$N = \frac{e^2}{\mu} v^2 = \frac{v^2}{\kappa} \tag{4.118}$$

so that the self-dual potential (4.110) reduces to

$$V = \frac{e^2}{2} \left(|\phi|^2 - v^2 \right)^2 \tag{4.119}$$

which is the quartic self-dual potential (1.8) of the abelian Higgs model [196,24,137]. Furthermore, in the nonrelativistic limit the Lagrange density (4.109) reduces to the Lagrange density (2.122) of the nonrelativistic Maxwell-Chern-Simons model described in Section II H.

The self-duality of the full relativistic Maxwell-Chern-Simons system with Lagrange density (4.109) can be obtained from a Bogomol'nyi style lower bound for the energy. First we record the Euler-Lagrange equations of motion

$$D_\mu D^\mu \phi = \frac{\partial V}{\partial \phi^*} = \phi \left(N - \frac{e^2}{\mu} v^2 \right)^2 + e^2 \phi \left(|\phi|^2 - \frac{\mu}{e^2} N \right) \qquad (4.120a)$$

$$\partial_\mu \partial^\mu N = e^2 \frac{\partial V}{\partial N} = 2e^2 |\phi|^2 \left(N - \frac{e^2}{\mu} v^2 \right) - \mu e^2 \left(|\phi|^2 - \frac{\mu}{e^2} N \right)$$

$$(4.120b)$$

$$\frac{1}{e^2} \partial_\mu F^{\mu\nu} + \frac{\mu}{2e^2} \epsilon^{\nu\alpha\beta} F_{\alpha\beta} = J^\nu \qquad (4.120c)$$

where J^ν is the relativistic current (4.5) for the complex scalar field ϕ. The zeroth component of the gauge equation (4.120c) is the Gauss law constraint

$$\frac{1}{e^2} \vec{\nabla} \cdot \vec{E} = \kappa B - J^0 \qquad (4.121)$$

Thus, the usual Chern-Simons charge-flux relation holds

$$\Phi \equiv \int d^2 x B = \frac{1}{\kappa} \int d^2 x J^0 \qquad (4.122)$$

The energy density for the system (4.109) is

$$\mathcal{E} = \frac{1}{2e^2} \vec{E}^2 + \frac{1}{2e^2} B^2 + |D_0 \phi|^2 + |\vec{D}\phi|^2 + \frac{1}{2e^2} (\partial_0 N)^2$$

$$+ \frac{1}{2e^2} \left(\vec{\nabla} N \right)^2 + V \qquad (4.123)$$

Using the Gauss law (4.121) together with the factorization identity (4.10) we can rewrite the energy density as

$$\mathcal{E} = \frac{1}{2e^2}\left(\vec{E} - \vec{\nabla}N\right)^2 - \frac{1}{e^2}N\vec{\nabla}\cdot\vec{E} + |D_0\phi|^2 + \frac{1}{2e^2}\left(\partial_0 N\right)^2$$

$$+|D_-\phi|^2 + B|\phi|^2 + \frac{1}{2e^2}B^2 + V$$

$$= \frac{1}{2e^2}\left(\vec{E} - \vec{\nabla}N\right)^2 + \left|D_0\phi - iN\phi + i\frac{v^2}{\kappa}\phi\right|^2 + \frac{1}{2e^2}\left(\partial_0 N\right)^2$$

$$+\frac{1}{2e^2}\left(B + e^2|\phi|^2 - e^2\kappa N\right)^2 + |D_-\phi|^2 + V - N^2|\phi|^2$$

$$-\frac{e^2}{2}|\phi|^4 - \frac{e^2\kappa^2}{2}N^2 + e^2\kappa N|\phi|^2 - \frac{v^4}{\kappa^2}|\phi|^2 + \frac{2v^2}{\kappa}N|\phi|^2$$

$$-i\frac{v^2}{\kappa}\left(\phi^* D^0\phi - \left(D^0\phi\right)^*\phi\right) \tag{4.124}$$

Thus, with the potential $V(|\phi|, N)$ chosen to be of the self-dual form (4.110) the energy is bounded below by

$$E \geq \frac{v^2}{\kappa}\int d^2x J^0 = v^2\Phi \tag{4.125}$$

This lower bound is saturated by fields satisfying the Gauss law (4.121) together with

$$D_-\phi = 0 \tag{4.126a}$$

$$B = -e^2|\phi|^2 + e^2\kappa N \tag{4.126b}$$

$$D_0\phi = -i\frac{v^2}{\kappa}\phi + iN\phi \tag{4.126c}$$

$$\vec{E} - \vec{\nabla}N = 0 \tag{4.126d}$$

$$\partial_0 N = 0 \tag{4.126e}$$

117

The last two of these equations are satisfied with static fields ($\partial_0 N = 0 = \partial_0 A_i$) and with N identified with A_0, in which case the third equation is solved by

$$\phi = e^{-iv^2 t/\kappa} \phi_{\text{static}} \qquad (4.127)$$

as in (4.20). Taking the divergence of (4.126d) and using the Gauss law (4.121) we find

$$\nabla^2 N = -e^4 \kappa |\phi|^2 + \kappa^2 e^4 N + 2e^2 |\phi|^2 N - 2\frac{e^2 v^2}{\kappa} |\phi|^2 \qquad (4.128)$$

which is just the static equation of motion (4.120b) for the neutral scalar field N. Thus the self-duality equations (4.126) are compatible with the Euler-Lagrange equations of motion (4.120). This system may be generalized to one with a nonabelian gauge symmetry and may be understood in terms of an extended supersymmetric model [166,142]. We also note here that the relativistic self-dual Chern-Simons models discussed in this Chapter have been generalized to include additional magnetic moment interactions [231,172,8], in which case the self-dual potential is *quadratic*, rather than sextic. Also, the relativistic self-dual Chern-Simons system exhibits many novel properties, with important applications to symmetry breaking, when considered in a uniform background field [171].

V. NONABELIAN RELATIVISTIC MODEL

This Chapter extends the relativistic self-dual Chern-Simons model to include nonabelian local gauge symmetry. This extension introduces many new features related to the interplay of the self-duality with the algebraic structure of the fields [169,170,41,62]. There is once again a Bogomol'nyi lower bound on the energy which is saturated by solutions to a set of first-order self-duality equations. This is true for a general matter coupling, but (as in the corresponding nonrelativistic models) the self-duality equations exhibit additional special properties with adjoint matter coupling, for which one can define matter and gauge fields in the same (arbitrary) representation of the gauge Lie algebra. In this case, the self-dual potential has an intricate vacuum structure, with vacua that are classified by embeddings of $SU(2)$ into the gauge algebra [140,64,65]. In these vacua, there are massive excitations for both gauge and scalar fields, and the masses follow unusual patterns which reflect the self-dual nature of the system and its associated $N = 2$ supersymmetry. In each vacuum each mass is paired, sometimes as complex scalar fields but in other cases as pairs of mass-degenerate real gauge and real scalar fields. These real fields have masses given by a simple universal mass formula in terms of the *exponents* of the gauge algebra. Moreover, the corresponding self-dual Chern-Simons equations constitute a deformation of the classical Toda system encountered in the nonrelativistic nonabelian theory in Chapter 3.

119

A. Relativistic Self-Dual Chern–Simons Equations: General Matter Coupling

The relativistic self-dual Chern-Simons system discussed in Chapter 4 may be generalized from an abelian theory to a nonabelian theory [169,170]. We consider now a multiplet Φ of complex scalar fields which transform under nonabelian local gauge transformations according to some definite representation $\mathcal{R}$ of the gauge algebra $\mathcal{G}$. The Lagrange density (4.1) becomes

$$\mathcal{L} = \frac{\kappa}{2}\mathcal{L}_{\mathrm{CS}} - (D^\mu \Phi)^\dagger (D_\mu \Phi) - V(\Phi^\dagger \Phi) \tag{5.1}$$

where $\mathcal{L}_{\mathrm{CS}}$ is the nonabelian Chern-Simons Lagrangian in (3.4) and V is the scalar potential

$$V = \frac{1}{\kappa^2} \left(\left(\Phi^\dagger \mathcal{T}^a \Phi\right) \mathcal{T}^a \Phi + v^2 \Phi \right)^\dagger \left(\left(\Phi^\dagger \mathcal{T}^a \Phi\right) \mathcal{T}^a \Phi + v^2 \Phi \right) \tag{5.2}$$

Here, the $\mathcal{T}^a$ are the (antihermitean) generators of the gauge algebra in the matter representation, and the matter and gauge fields are minimally coupled through the covariant derivative

$$D_\mu \Phi = \partial_\mu \Phi + A_\mu^a \mathcal{T}^a \Phi \tag{5.3}$$

As in the abelian case, the parameter v^2 appearing in the potential (5.2) plays the role of a mass scale. The Euler-Lagrange equations for this system are

$$D_\mu D^\mu \Phi = \frac{\partial V}{\partial \Phi^\dagger} \tag{5.4a}$$

$$F_{\mu\nu}^a = -\frac{1}{\kappa}\epsilon_{\mu\nu\rho} J^{a\,\rho} \tag{5.4b}$$

where $F_{\mu\nu}^a$ is the nonabelian gauge curvature and $J^{a\,\mu}$ is the gauge current

120

$$J_\mu^a = - \left(\Phi^\dagger T^a D_\mu \Phi - (D_\mu \Phi)^\dagger T^a \Phi \right) \tag{5.5}$$

which satisfies the covariant conservation equation

$$\partial_\mu J^{a\,\mu} + f^{abc} A_\mu^b J^{c\,\mu} = 0 \tag{5.6}$$

In addition to the gauge current $J^{a\,\mu}$ there is an abelian current Q^μ, corresponding to the global $U(1)$ symmetry of the Lagrange density (5.1)

$$Q_\mu = -i \left(\Phi^\dagger D_\mu \Phi - (D_\mu \Phi)^\dagger \Phi \right) \tag{5.7}$$

which satisfies the ordinary continuity equation

$$\partial_\mu Q^\mu = 0 \tag{5.8}$$

This global $U(1)$ symmetry is crucial for the association of the self-dual theory with an $N = 2$ supersymmetric model [169,120,95]. The energy density corresponding to the Lagrange density (5.1) is

$$\mathcal{E} = (D_0 \Phi)^\dagger D_0 \Phi + (D_i \Phi)^\dagger D_i \Phi + V \tag{5.9}$$

supplemented by the Gauss law constraint

$$F_{12}^a = \frac{1}{\kappa} J^{a0} = -\frac{1}{\kappa} \left(\Phi^\dagger T^a D^0 \Phi - \left(D^0 \Phi\right)^\dagger T^a \Phi \right) \tag{5.10}$$

which is the zeroth component of the gauge field equations of motion (5.4b).

To find self-dual solutions which minimize the energy, we use the nonabelian factorization identity (3.23), together with the Gauss law constraint (5.10), to write

$$(D_i \Phi)^\dagger D_i \Phi = (D_- \Phi)^\dagger D_- \Phi$$

$$-\frac{i}{\kappa} \left(\Phi^\dagger T^a \Phi \right) \left(\Phi^\dagger T^a D_0 \Phi - (D_0 \Phi)^\dagger T^a \Phi \right) \tag{5.11}$$

Now write

$$(D_0\Phi)^\dagger D_0\Phi = \left| D_0\Phi + \frac{i}{\kappa} \left(\left(\Phi^\dagger \mathcal{T}^a \Phi\right) \mathcal{T}^a \Phi + v^2\Phi \right) \right|^2$$

$$- \frac{i}{\kappa} \left(\left(\Phi^\dagger \mathcal{T}^a \Phi\right) \mathcal{T}^a \Phi + v^2\Phi \right)^\dagger D_0\Phi$$

$$+ \frac{i}{\kappa} (D_0\Phi)^\dagger \left(\left(\Phi^\dagger \mathcal{T}^a \Phi\right) \mathcal{T}^a \Phi + v^2\Phi \right)$$

$$- \frac{1}{\kappa^2} \left| \left(\Phi^\dagger \mathcal{T}^a \Phi\right) \mathcal{T}^a \Phi + v^2\Phi \right|^2 \tag{5.12}$$

where $|M|^2$ denotes $M^\dagger M$. The final term in this expression (5.12) is recognized as (minus) the potential V defined in (5.2), and so the energy density (5.9) can be expressed as

$$\mathcal{E} = \left| D_0\Phi + \frac{i}{\kappa} \left(\left(\Phi^\dagger \mathcal{T}^a \Phi\right) \mathcal{T}^a \Phi + v^2\Phi \right) \right|^2$$

$$+ (D_-\Phi)^\dagger D_-\Phi + \frac{iv^2}{\kappa} \left(\Phi^\dagger (D_0\Phi) - (D_0\Phi)^\dagger \Phi \right) \tag{5.13}$$

The first two terms in (5.13) are manifestly positive and the third gives a lower bound for the energy density, which may be written in terms of the time component, Q^0, of the abelian relativistic current defined in (5.7):

$$\mathcal{E} \geq \frac{v^2}{\kappa} Q^0 \tag{5.14}$$

(Note that we have assumed that κ is positive - if κ were negative then $\mathcal{E}$ should be factorized using $D_+\Phi$ instead of $D_-\Phi$, just as in the abelian case.)

This lower bound (5.14) is saturated when the following two conditions (each first order in spacetime derivatives) hold:

$$D_-\Phi = 0 \tag{5.15a}$$

122

$$D_0 \Phi = -\frac{i}{\kappa} \left(\left(\Phi^\dagger \mathcal{T}^a \Phi \right) \mathcal{T}^a \Phi + v^2 \Phi \right) \tag{5.15b}$$

The consistency condition of these two equations states that

$$(D_0 D_- - D_- D_0) \Phi = \frac{i}{\kappa} \left((D_+\Phi)^\dagger \mathcal{T}^a \Phi \right) \mathcal{T}^a \Phi \tag{5.16}$$

But

$$(D_0 D_- - D_- D_0) \Phi \equiv F_{0-}^a \mathcal{T}^a \Phi \tag{5.17}$$

so we see that

$$F_{0-}^a = \frac{i}{\kappa} \left((D_+\Phi)^\dagger \mathcal{T}^a \Phi \right) \tag{5.18}$$

Furthermore, using the self-dual ansatz (5.15a), the spatial components of the gauge current (5.5) simplify to

$$J_-^a \equiv J_1^a - i J_2^a = (D_+\Phi)^\dagger \mathcal{T}^a \Phi \tag{5.19}$$

so that (5.18) is just the spatial part of the gauge field equation of motion (5.4b). The temporal part (5.10) of the gauge field equation may be re-expressed using equation (5.15b) in a form not involving explicit time derivatives. We thus arrive at the *relativistic self-dual Chern-Simons equations*:

$$D_- \Phi = 0 \tag{5.20a}$$

$$F_{+-}^a = -\frac{2}{\kappa^2} \left(\left(\Phi^\dagger \mathcal{T}^b \Phi \right) \Phi^\dagger \{\mathcal{T}^a, \mathcal{T}^b\} \Phi + 2v^2 \left(\Phi^\dagger \mathcal{T}^a \Phi \right) \right) \tag{5.20b}$$

In the abelian limit in which we replace the (antihermitean) generators $\mathcal{T}^a$ by i these equations reduce to the abelian relativistic self-dual Chern-Simons equations (4.19) (recall that $F_{+-} = -2i F_{12}$). Furthermore, the nonrelativistic limit of these nonabelian relativistic

123

self-duality equations (5.20) proceeds in exactly the same manner as for the abelian theory (see Section IV D). Unlike the nonrelativistic case, the solutions to the relativistic self-duality equations (5.20) are not necessarily static, although their time dependence is simple owing to the relation (5.15b). We can write the solution as

$$\Phi = e^{-imx^0} \Phi_{\text{static}} \tag{5.21}$$

with

$$A_0^a = -\frac{i}{\kappa} \left(\Phi^\dagger \mathcal{T}^a \Phi \right) \tag{5.22}$$

To check the consistency of the self-duality equations (5.20) with the matter Euler-Lagrange equation of motion (5.4a) we note that from (5.15b) and (5.22)

$$D_0 D_0 \Phi = -\frac{1}{\kappa^2} \left[\frac{1}{2} \left(\Phi^\dagger \mathcal{T}^a \Phi \right) \left(\Phi^\dagger \mathcal{T}^b \Phi \right) \{ \mathcal{T}^a, \mathcal{T}^b \} \Phi \right.$$

$$\left. + 2v^2 \left(\Phi^\dagger \mathcal{T}^a \Phi \right) \mathcal{T}^a \Phi + v^4 \Phi \right] \tag{5.23}$$

Therefore

$$D_\mu D^\mu \Phi = -D_0 D_0 \Phi + D_i D_i \Phi$$

$$= -D_0 D_0 \Phi + D_+ D_- \Phi + i F_{12}^a \mathcal{T}^a \Phi$$

$$= \frac{1}{\kappa^2} \left[\frac{1}{2} \left(\Phi^\dagger \mathcal{T}^a \Phi \right) \left(\Phi^\dagger \mathcal{T}^b \Phi \right) \{ \mathcal{T}^a, \mathcal{T}^b \} \Phi \right.$$

$$+ \left(\Phi^\dagger \mathcal{T}^a \Phi \right) \left(\Phi^\dagger \{ \mathcal{T}^a, \mathcal{T}^b \} \Phi \right) \mathcal{T}^a \Phi$$

$$\left. + 4v^2 \left(\Phi^\dagger \mathcal{T}^a \Phi \right) \mathcal{T}^a \Phi + v^4 \Phi \right]$$

$$= \frac{\partial V}{\partial \Phi^\dagger} \tag{5.24}$$

where the potential V takes its self-dual form (5.2).

In the relativistic theory it is possible to have nontrivial solutions for Φ while still having $F^a_{+-} = 0$. These solutions correspond to minima of the potential (5.2) and are constant fields $\Phi_{(0)}$ satisfying the nonlinear algebraic relation

$$\left(\Phi^\dagger_{(0)} T^a \Phi_{(0)}\right) T^a \Phi_{(0)} = -v^2 \Phi_{(0)} \tag{5.25}$$

From (5.15b) we see that

$$D_0 \Phi_{(0)} = 0 \tag{5.26}$$

and so the relativistic abelian charge density Q^0 in (5.7) vanishes. Hence, from (5.14) the energy of these configurations is zero (as it should be).

B. Relativistic Self-Dual Chern–Simons Equations: Adjoint Matter Coupling

The nonabelian relativistic self-dual Chern-Simons system has additional special features with adjoint matter coupling [170,62]. With the matter field Φ in the adjoint representation (which has the same dimension as the gauge algebra) one can take the generators to be

$$(\mathcal{T}^a)_{bc} = f^{abc} \tag{5.27}$$

where the f^{abc} are the structure constants (3.5) of the gauge Lie algebra. Then we can define Lie algebra valued fields

$$\phi = \Phi^a T^a$$

$$A_\mu = A^a_\mu T^a \tag{5.28}$$

where the T^a are (antihermitean) generators in *any* representation of the algebra. With the generators normalized as in (3.29) the Lagrange density (5.1) of the nonabelian relativistic self-dual Chern-Simons system becomes

$$\mathcal{L} = -\frac{\kappa}{2}\mathcal{L}_{CS} - tr\left((D_\mu\phi)^\dagger D^\mu\phi\right) - V \qquad (5.29)$$

where $D_\mu\phi = \partial_\mu\phi + [A_\mu, \phi]$, and the nonabelian Chern-Simons Lagrange density $\mathcal{L}_{CS}$ is given by (3.31). The self-dual scalar field potential V is

$$V = \frac{1}{\kappa^2}tr\left(\left([\,[\,\phi,\phi^\dagger\,],\phi\,] - v^2\phi\right)^\dagger \left([\,[\,\phi,\phi^\dagger\,],\phi\,] - v^2\phi\right)\right). \qquad (5.30)$$

The Euler-Lagrange equations of motion obtained from the Lagrange density (5.29) are:

$$D_\mu D^\mu\phi = \frac{\partial V}{\partial\phi^\dagger} \qquad (5.31a)$$

$$F_{\mu\nu} = \frac{i}{\kappa}\epsilon_{\mu\nu\rho}J^\rho \qquad (5.31b)$$

where J^μ is the relativistic nonabelian gauge current

$$J^\mu \equiv -i\left([\,\phi^\dagger, D^\mu\phi\,] - [\,(D^\mu\phi)^\dagger, \phi\,]\right) \qquad (5.32)$$

which is covariantly conserved

$$D_\mu J^\mu = 0 \qquad (5.33)$$

This system also has an abelian current, Q_μ, corresponding to the global $U(1)$ invariance of the Lagrange density (5.29)

$$Q_\mu = -i\,tr\left(\phi^\dagger D_\mu\phi - (D_\mu\phi)^\dagger\phi\right), \qquad (5.34)$$

which is ordinarily conserved

$$\partial_\mu Q^\mu = 0 \tag{5.35}$$

The energy density corresponding to the Lagrange density (4.1) is

$$\mathcal{H} = tr\left((D_0\phi)^\dagger D_0\phi\right) + tr\left((D_i\phi)^\dagger D_i\phi\right) + V\left(\phi, \phi^\dagger\right), \tag{5.36}$$

supplemented by the Gauss law constraint

$$F_{12} = -\frac{i}{\kappa}J^0 = \frac{1}{\kappa}\left([\phi^\dagger, D_0\phi] - [(D_0\phi)^\dagger, \phi]\right) \tag{5.37}$$

To find self-dual solutions which minimize the energy, we re-express the energy density in a factorized form, as in (5.13). Using the identity (3.23) together with the Gauss law constraint (5.37), we can write (modulo unimportant surface terms)

$$tr\left((D_i\phi)^\dagger D_i\phi\right) = tr\left((D_-\phi)^\dagger D_-\phi\right)$$

$$+\frac{i}{\kappa}tr\left(([[\phi, \phi^\dagger], \phi])^\dagger D_0\phi - [[\phi, \phi^\dagger], \phi](D_0\phi)^\dagger\right) \tag{5.38}$$

The terms involving $D_0\phi$ may be cancelled in the energy density (5.36) by a term from $tr\left((D_0\phi)^\dagger D_0\phi\right)$ if we write

$$tr\left((D_0\phi)^\dagger D_0\phi\right) = tr\left(\left|D_0\phi - \frac{i}{\kappa}\left([[\phi, \phi^\dagger], \phi] - v^2\phi\right)\right|^2\right)$$

$$-\frac{i}{\kappa}tr\left(\left([[\phi, \phi^\dagger], \phi] - v^2\phi\right)^\dagger D_0\phi\right)$$

$$+\frac{i}{\kappa}tr\left(\left([[\phi, \phi^\dagger], \phi] - v^2\phi\right)(D_0\phi)^\dagger\right)$$

$$-\frac{1}{\kappa^2}tr\left(\left|[[\phi, \phi^\dagger], \phi] - v^2\phi\right|^2\right) \tag{5.39}$$

The final term in this expression (5.39) is recognized as (minus) the self-dual potential V defined in (5.30), and so the energy density (5.36) can be expressed as

$$\mathcal{E} = tr \left(\left| D_0 \phi - \frac{i}{\kappa} \left([[\phi, \phi^\dagger], \phi] - v^2 \phi \right) \right|^2 \right)$$

$$+ tr \left((D_- \phi)^\dagger D_- \phi \right) + \frac{iv^2}{\kappa} tr \left(\phi^\dagger (D_0 \phi) - (D_0 \phi)^\dagger \phi \right) \tag{5.40}$$

The first two terms in (5.40) are manifestly positive and the third gives a lower bound for the energy density, which may be written in terms of the time component, Q^0, of the abelian relativistic current defined in (5.34):

$$\mathcal{E} \geq \frac{v^2}{\kappa} Q^0 \tag{5.41}$$

This lower bound (5.41) is saturated when the following two conditions (each first order in spacetime derivatives) hold:

$$D_- \phi = 0 \tag{5.42a}$$

$$D_0 \phi = \frac{i}{\kappa} \left([[\phi, \phi^\dagger], \phi] - v^2 \phi \right) \tag{5.42b}$$

The consistency condition of these two equations states that

$$(D_0 D_- - D_- D_0) \phi \equiv [F_{0-}, \phi]$$

$$= -\frac{i}{\kappa} [[\phi, (D_+ \phi)^\dagger], \phi]$$

$$= \frac{1}{\kappa} [J_-, \phi] \tag{5.43}$$

which expresses the gauge field Euler-Lagrange equation of motion

$$F_{0-} = \frac{1}{\kappa} J_-$$

$$= -\frac{i}{\kappa} [\phi, (D_+ \phi)^\dagger] \tag{5.44}$$

for the spatial components of the self-dual current. The other gauge field equation, the Gauss law constraint (5.37), may be re- expressed using equation (5.42b) in a form not involving explicit time derivatives. We thus arrive at the *relativistic self-dual Chern-Simons equations*:

$$D_-\phi = 0 \tag{5.45a}$$

$$F_{+-} = \frac{4}{\kappa^2}[v^2\phi - [[\phi, \phi^\dagger], \phi], \phi^\dagger] \tag{5.45b}$$

At the self-dual point, we can use equation (5.42b) to express the energy density as

$$\mathcal{E}_{SD} = \frac{2v^2}{\kappa^2} tr\left(\phi^\dagger\left(v^2\phi - [[\phi, \phi^\dagger], \phi]\right)\right) \tag{5.46}$$

Zero energy self-dual solutions ϕ are gauge equivalent to solutions of the *algebraic* equation

$$[[\phi, \phi^\dagger], \phi] = v^2\phi. \tag{5.47}$$

Solutions of this equation also correspond to the minima of the potential (5.30), and these potential minima are clearly degenerate. Then a class of solutions to the self-duality equations (5.45) is given by the following zero energy solutions of the Euler-Lagrange equations:

$$\phi = g^{-1}\phi_{(0)}g$$

$$A_\pm = g^{-1}\partial_\pm g$$

$$A_0 = g^{-1}\partial_0 g \tag{5.48}$$

where $\phi_{(0)}$ is any solution of (5.47), and $g = g(\vec{x}, t)$ takes values in the gauge group. It is clear that these solutions satisfy $D_0\phi = 0$, $D_-\phi = 0$, $F_{+-} = 0$, as well as the algebraic equation (5.47), which implies that they are self-dual, and that they have zero magnetic field and zero charge density. While this class of solutions may look somewhat trivial, it is still important because the solutions, $\phi_{(0)}$, of the algebraic equation (5.47) classify the minima of the potential V, and the finite *nonzero* energy solutions of the self-duality equations must be gauge equivalent to such a solution at infinity:

$$\phi \to g^{-1}\phi_{(0)}g \qquad \text{as} \quad r \to \infty \tag{5.49}$$

Before considering these vacuum configurations, we check explicitly the consistency of the self-duality equations (5.45) with the Euler-Lagrange equation of motion (5.31). Note that

$$D_\mu D^\mu \phi = -D_0 D_0 \phi + D_i D_i \phi$$

$$= -D_0 D_0 \phi + D_+ D_- \phi + i[F_{12}, \phi] \tag{5.50}$$

For self-dual solutions $D_- \phi = 0$, and using the self-duality equation for $D_0 \phi$ we find that

$$i[F_{12}, \phi] = \frac{2}{\kappa^2}[[\phi, [\phi^\dagger, [\phi, \phi^\dagger]]], \phi] + \frac{2v^2}{\kappa^2}[\phi, [\phi, \phi^\dagger]] \tag{5.51}$$

and

$$-D_0 D_0 \phi = \frac{v^4}{\kappa^2}\phi + \frac{2v^2}{\kappa^2}[\phi, [\phi, \phi^\dagger]] + \frac{1}{\kappa^2}[[\phi, [\phi, \phi^\dagger]], [\phi, \phi^\dagger]] \tag{5.52}$$

Therefore,

$$D_\mu D^\mu \phi = \frac{v^4}{\kappa^2}\phi + \frac{4v^2}{\kappa^2}[\phi, [\phi, \phi^\dagger]]$$

$$+ \frac{1}{\kappa^2}\left([[\phi, [\phi, \phi^\dagger]], [\phi, \phi^\dagger]] + 2[[\phi, [\phi^\dagger, [\phi, \phi^\dagger]]], \phi]\right) \tag{5.53}$$

It is a straightforward matter to verify that (5.53) does indeed yield the correct charged scalar field Euler-Lagrange equation of motion (5.31a) with the self-dual potential V given by equation (5.30).

C. Classification of Minima

The sixth order self-dual potential (5.30) has degenerate minima given by fields $\phi_{(0)}$ which solve

$$[[\phi, \phi^\dagger], \phi] = \phi \qquad (5.54)$$

where for convenience a factor of $|v|$ has been absorbed into the field ϕ. We recognize the condition (5.54) as the $SU(2)$ commutation relation [140,64]. For ϕ taking values in a general gauge algebra, finding the solutions to (5.54) is the classic Dynkin problem [70,157] of embedding $SU(2)$ into a general Lie algebra. It is interesting to note that this type of embedding problem also plays a significant role in the theory of spherically symmetric magnetic monopoles [179,100] and the Toda molecule equations [93].

Once again, for ease of presentation we consider $SU(N)$ - the generalization to other gauge algebras is relatively straightforward. It is clear that in order to satisfy (5.54) for a general gauge algebra, $\phi = \phi_{(0)}$ must be a linear combination of the step operators for the *positive* roots of the algebra. Further, since we have the freedom of global gauge invariance, we can choose representative gauge inequivalent solutions $\phi_{(0)}$ to be linear combinations of the step operators of the positive *simple* roots. It is therefore convenient to work in the Chevalley basis (3.51) for the gauge algebra. Expand $\phi_{(0)}$ in terms of the positive simple root step operators as:

$$\phi_{(0)} = \sum_{a=1}^{N-1} \phi_{(0)}^a E_a \qquad (5.55)$$

Then $[\phi_{(0)}, \phi_{(0)}^\dagger]$ is diagonal,

$$[\phi_{(0)}, \phi_{(0)}^\dagger] = \sum_{a=1}^{N-1} |\phi_{(0)}^a|^2 H_a. \qquad (5.56)$$

The Chevalley basis commutation relations (3.51) then imply that

$$[[\phi_{(0)}, \phi_{(0)}^\dagger], \phi_{(0)}] = \sum_{a=1}^{N-1} \sum_{b=1}^{N-1} |\phi_{(0)}^a|^2 \phi_{(0)}^b C_{ba} E_b \tag{5.57}$$

which, like $\phi_{(0)}$, is once again a linear combination of just the simple root step operators. Thus, for suitable choices of the coefficients $\phi_{(0)}^a$, it is *possible* for the $SU(N)$ algebra element $\phi_{(0)}$ to satisfy the $SU(2)$ commutation relation $[[\phi, \phi^\dagger], \phi] = \phi$.

For example, one can always choose $\phi_{(0)}$ proportional to a *single* step operator, which by global gauge invariance can always be taken to be E_1 :

$$\phi_{(0)} = \frac{1}{\sqrt{2}} E_1 \tag{5.58}$$

In the other extreme, the $SU(N)$ "maximal embedding" case, with *all* $N-1$ step operators involved in the expansion (5.55), the solution for $\phi_{(0)}$ is[5]

$$\phi_{(0)} = \frac{1}{\sqrt{2}} \sum_{a=1}^{N-1} \sqrt{a(N-a)} \, E_a \tag{5.59}$$

All other solutions for $\phi_{(0)}$, intermediate between the two extremes (5.58) and (5.59), can be generated by the following systematic procedure. If one of the simple root step operators, say E_b, is omitted from the summation in (5.55) then this effectively decouples the $E_{\pm a}$'s with $a < b$ from those with $a > b$. Then the coefficients for the $(b-1)$ step operators E_a with $a < b$ are just those for the maximal

---

[5]In general, the squares of the coefficients for the maximal embedding case are the coefficients, in the simple root basis, of (one half times) the sum of *all* positive roots of the algebra [64].

embedding (see equation (5.59)) in $SU(b)$, and the coefficients for the $(N-b-1)$ E_a's with $a > b$ are those for the maximal embedding in $SU(N-b)$:

$$\phi_{(0)} = \frac{1}{\sqrt{2}} \sum_{a=1}^{b-1} \sqrt{a(b-a)} E_a + \frac{1}{\sqrt{2}} \sum_{a=b+1}^{N-1} \sqrt{a(N-b-a)} E_a \qquad (5.60)$$

Diagrammatically, we can represent the maximal embedding case (5.59) with the Dynkin diagram of $SU(N)$:

$$\underbrace{o - o - o - \ldots - o - o}_{N-1} \qquad (5.61)$$

which shows the $N-1$ simple roots of the algebra, each connected to its nearest neighbours by a single line. Omitting the bth simple root step operator from the sum in (5.55) can be conveniently represented as breaking the Dynkin diagram in two by deleting the bth dot:

$$\underbrace{o - o - \ldots - o}_{b-1} - \times - \underbrace{o - \ldots - o}_{N-b-1} \qquad (5.62)$$

With this deletion of the bth dot, the $SU(N)$ Dynkin diagram breaks into the Dynkin diagram for $SU(b)$ and that for $SU(N-b)$. Since the remaining simple root step operators decouple into a Chevalley basis for $SU(b)$ and another for $SU(N-b)$, the coefficients required for the summation over the first $b-1$ step operators are just those given in (5.59) for the maximal embedding in $SU(b)$, while the coefficients for the summation over the last $N-b-1$ step operators are given by the maximal embedding for $SU(N-b)$, as indicated in (5.60).

It is clear that this process may be repeated with further roots being deleted from the Dynkin diagram, thereby subdividing the original $SU(N)$ Dynkin diagram, with its $N-1$ consecutively linked dots, into subdiagrams of $\leq N-1$ consecutively linked dots. The final diagram, with M deletions made, can be characterized, up to

133

gauge equivalence, by the $M+1$ lengths of the remaining consecutive strings of dots. A simple counting argument shows that the total number of ways of doing this (including the case where *all* dots are deleted, which corresponds to the trivial solution $\phi_{(0)} = 0$) is given by the number, $p(N)$, of (unrestricted) partitions of N.

The $SU(4)$ case is sufficient to illustrate this procedure. There are 5 partitions of 4, and they correspond to the following solutions for $\phi_{(0)}$:

$$o - o - o \qquad\qquad \phi_{(0)} = \frac{1}{\sqrt{2}}\left(\sqrt{3}E_1 + 2E_2 + \sqrt{3}E_3\right)$$

$$o - o - \times \qquad\qquad \phi_{(0)} = E_1 + E_2$$

$$o - \times - o \qquad\qquad \phi_{(0)} = \frac{1}{\sqrt{2}}E_1 + \frac{1}{\sqrt{2}}E_3$$

$$o - \times - \times \qquad\qquad \phi_{(0)} = \frac{1}{\sqrt{2}}E_1$$

$$\times - \times - \times \qquad\qquad \phi_{(0)} = 0 \qquad\qquad\qquad (5.63)$$

Thus we have a simple constructive procedure, and a correspondingly simple labelling notation, for finding all $p(N)$ gauge inequivalent solutions $\phi_{(0)}$ to the algebraic embedding condition (5.54). Recall that each such $\phi_{(0)}$ characterizes a distinct minimum of the potential V, as well as a class of zero energy solutions to the selfduality equations (5.45).

Since each vacuum solution $\phi_{(0)}$ corresponds to an embedding of $SU(2)$ into $SU(N)$, an alternative shorthand for labelling the different vacua consists of listing the block diagonal spin content of the $SU(2)$ Cartan subagebra element $[\phi_{(0)}, \phi_{(0)}^\dagger] \sim J_3$. For example, consider the matter fields ϕ taking values in the $N \times N$ defining representation. Then, for each vacuum solution, $[\phi_{(0)}, \phi_{(0)}^\dagger]$ takes the $N \times N$ diagonal sub-blocked form:

$$[\phi_{(0)}, \phi^\dagger_{(0)}] = \begin{pmatrix} j_1 & & & & & & & & \\ & \ddots & & & & & & & \\ & & -j_1 & & & & & & \\ & & & j_2 & & & & & \\ & & & & \ddots & & & & \\ & & & & & -j_2 & & & \\ & & & & & & \ddots & & \\ & & & & & & & j_M & \\ & & & & & & & & \ddots & \\ & & & & & & & & & -j_M \end{pmatrix}$$

Each spin j sub-block has dimension $2j + 1$, and so it is therefore natural to associate this particular $\phi_{(0)}$ with the following partition of N :

$$N = (2j_1 + 1) + (2j_2 + 1) + \ldots + (2j_M + 1) \qquad (5.64)$$

For example, the $SU(4)$ solutions listed in (5.63) may be labelled by the partitions $4, 3+1, 2+2, 2+1+1$, and $1+1+1+1$, respectively.

D. Vacuum Mass Spectra

Having classified the gauge inequivalent vacua of the self-dual potential V, we now determine the spectrum of massive excitations in each vacuum. In the abelian model there is only one nontrivial vacuum, and in this broken vacuum the massive gauge excitation and the remaining real massive scalar field are degenerate in mass [113,127,129]. This degeneracy of the gauge and scalar masses in the broken vacuum is also a feature of the $2+1$ dimensional Abelian

Higgs model [24]. In the nonabelian models considered here the situation is considerably more complicated, due to the presence of many fields and also due to the many different gauge inequivalent vacua. Nevertheless, we shall see that an analogous mass degeneracy pattern exists, reflecting the self-dual character of the potential (5.30).

In the unbroken vacuum, with $\phi_{(0)} = 0$, there are $N^2 - 1$ complex scalar fields, each with mass

$$m = \frac{v^2}{\kappa} \qquad (5.65)$$

In a broken vacuum, where $\phi_{(0)} \neq 0$, the scalar masses are determined by expanding the shifted potential $V(\phi + \phi_{(0)})$ to quadratic order in the field ϕ:

$$V(\phi + \phi_{(0)}) = \frac{v^4}{\kappa^2} tr \left(\left| [[\phi_{(0)}, \phi^\dagger], \phi_{(0)}] + [[\phi, \phi^\dagger_{(0)}], \phi_{(0)}] \right. \right.$$

$$\left. \left. + [[\phi_{(0)}, \phi^\dagger_{(0)}], \phi] - \phi \right|^2 \right) \qquad (5.66)$$

With the fields normalized appropriately, the masses are then given by the square roots of the eigenvalues of the $2(N^2 - 1) \times 2(N^2 - 1)$ mass matrix in (5.66).

In a broken vacuum, where $\phi_{(0)} \neq 0$, some of the original $2(N^2-1)$ massive scalar degrees of freedom are converted to massive gauge degrees of freedom via the Chern-Simons Higgs mechanism, as discussed in Section IV E. The gauge masses are determined by the quadratic gauge field Lagrange density

$$\mathcal{L}_{\text{quad}} = -\frac{\kappa}{2} \epsilon^{\mu\nu\rho} tr \left(A_\mu \partial_\nu A_\rho \right) - v^2 tr \left([A_\mu, \phi_{(0)}]^\dagger [A^\mu, \phi_{(0)}] \right) \qquad (5.67)$$

This quadratic Lagrange density describes massive gauge excitations with masses given by

$$m_{\text{gauge}} = \frac{2v^2}{\kappa} \lambda \qquad (5.68)$$

where the λ are the eigenvalues of the "gauge mass matrix"

$$tr\left([A_\mu, \phi_{(0)}]^\dagger [A^\mu, \phi_{(0)}]\right) \qquad (5.69)$$

Note that the gauge masses are determined by finding the eigenvalues (*not* the square roots of the eigenvalues) of the $(N^2 - 1) \times (N^2 - 1)$ mass matrix in (5.69). Thus the gauge masses are proportional to the same mass scale $\frac{v^2}{\kappa}$ as the scalar masses arising from (5.66).

This procedure of finding the eigenvalues of the scalar and gauge mass matrices, must be performed for each of the $p(N)$ gauge inequivalent minima $\phi_{(0)}$ of V. To illustrate the procedure, we first consider the case of $SU(2)$. For $SU(2)$ in the unbroken vacuum there are $6 (= 2 \times 3)$ real massive scalar fields each of mass $m = \frac{v^2}{\kappa}$. There is just one nontrivial vacuum, with

$$\phi_{(0)} = \frac{1}{\sqrt{2}} E_1 \qquad (5.70)$$

and the scalar masses in this broken vacuum are found by diagonalizing (5.66). The diagonalized form arises from the unitary decomposition

$$\phi = \frac{1}{\sqrt{2}} \chi E_1 + \xi E_{-1} \qquad (5.71)$$

where χ is a real field and ξ a complex field. Then the quadratic part of the scalar Lagrange density is

$$\mathcal{L}_{\text{quad}} = -\frac{1}{2} (\partial_\mu \chi)^2 - |\partial_\mu \xi|^2 - \frac{1}{2}\left(\frac{2v^2}{\kappa}\right)^2 \chi^2 - \left(\frac{2v^2}{\kappa}\right)^2 |\xi|^2 \qquad (5.72)$$

Thus, there is one real scalar field χ (the "Higgs" field) with mass

$$m_{\text{scalar}} = \frac{2v^2}{\kappa} = 2m \qquad (5.73)$$

and one complex scalar field ξ also of mass $2m$, giving a total of 3 massive scalar degrees of freedom.

The gauge fields are all massless in the unbroken vacuum, but in the broken vacuum (5.70) some components acquire mass by the Chern-Simons Higgs mechanism. Decomposing the gauge field as

$$A_\mu = \frac{i}{\sqrt{2}} A_\mu^3 H_1 + \frac{i}{\sqrt{2}} A_\mu^1 \sigma_1 + \frac{i}{\sqrt{2}} A_\mu^2 \sigma_2 \tag{5.74}$$

where $E_1 = (\sigma_1 + i\sigma_2)/2$, the quadratic gauge field Lagrange density (5.67) becomes

$$\mathcal{L}_{\text{quad}} = \frac{\kappa}{2} \epsilon^{\mu\nu\rho} A_\mu^a \partial_\nu A_\rho^a - v^2 (A_\mu^3)^2 - \frac{v^2}{2} (A_\mu^1)^2 - \frac{v^2}{2} (A_\mu^2)^2 \tag{5.75}$$

Therefore, in the broken vacuum (5.70) there is one real gauge field A_μ^3 of mass

$$m_{\text{gauge}} = \frac{2v^2}{\kappa} = 2m \tag{5.76}$$

and one "complex" gauge field $(A_\mu^1 \pm i A_\mu^2)/\sqrt{2}$ of mass m. By "complex" we simply mean two real fields of equal mass which naturally combine into a complex field. Thus, in the broken vacuum there are 3 massive gauge field degrees of freedom. Furthermore, the real gauge field mass (5.76) is equal to the real scalar field mass (5.73). The real scalar field χ comes from a field with the same algebraic decomposition as $\phi_{(0)}$, while the real gauge field comes from the diagonal gauge field component.

For larger gauge algebras this diagonalization process is more involved. The results for $SU(3)$ and $SU(4)$ are presented here in Tables I and II (see also [140,64]).

138

vacuum	gauge masses			
$\phi_{(0)}$	real fields	complex fields		
o — x	2	1/2	1/2	1
o — o	2 6	1	2	5

vacuum	scalar masses			
$\phi_{(0)}$	real fields	complex fields		
o — x	2	1	3/2	3/2	2
o — o	2 6	2	3	5

Table I. $SU(3)$ vacuum mass spectra, in units of the fundamental mass scale v^2/κ, for the inequivalent nontrivial minima $\phi_{(0)}$ of the potential V. Notice that for each vacuum the *total* number of massive degrees of freedom is equal to $2(N^2-1) = 16$, although the distribution between gauge and scalar fields is vacuum dependent.

A number of interesting observations can be made at this point, based on the evaluation of these mass spectra for the various vacua in $SU(N)$ for N up to 10.

(i) All masses, both gauge and scalar, are integer or half-odd-integer multiples of the fundamental mass scale $m = v^2/\kappa$. The fact that all the scalar masses are proportional to m is clear from the form of the potential V in (5.30). The fact that the gauge masses are multiples of the *same* mass scale depends on the fact that the Chern- Simons coupling parameter κ has been included in the overall normalization of the potential in (5.30). This is a direct consequence of the self-duality of the model.

vacuum	gauge masses	
$\phi_{(0)}$	real fields	complex fields
$o - \times - \times$	2	1/2 1/2 1/2 1/2 1
$o - \times - o$	2 2	1 1 1 1 2
$o - o - \times$	2 6	1 1 1 2 2 5
$o - o - o$	2 6 12	1 2 3 5 8 11

vacuum	scalar masses	
$\phi_{(0)}$	real fields	complex fields
$o - \times - \times$	2	1 1 1 1 3/2 3/2 3/2 3/2 2
$o - \times - o$	2 2	1 1 1 2 2 2 2 2
$o - o - \times$	2 6	1 2 2 2 2 3 5
$o - o - o$	2 6 12	2 3 4 5 8 11

Table II. $SU(4)$ vacuum mass spectra, in units of the fundamental mass scale $\frac{v^2}{\kappa}$, for the inequivalent nontrivial minima $\phi_{(0)}$ of the potential V. Notice that for each vacuum the *total* number of massive degrees of freedom is equal to $2(N^2-1) = 30$, although the distribution between gauge and scalar fields is vacuum dependent.

(ii) In each vacuum, the masses of the real scalar excitations are equal to the masses of the real gauge excitations, whereas this is not true of the complex scalar and gauge fields (by 'complex' gauge fields we simply mean those fields which naturally appear as complex combinations of the nonhermitean step operator generators). Indeed, in some vacua the *number* of complex scalar degrees of freedom and complex gauge degrees of freedom is not even the same. This will be discussed further below.

(iii) In each vacuum, each mass appears at least twice, and always an even number of times. For the complex fields this is a triviality, but for the real fields this is only true as a consequence of the feature mentioned in (ii). This pairing of the masses is a reflection of the $N = 2$ supersymmetry of the relativistic self-dual Chern-Simons systems [165,120].

(iv) While the distribution of masses between gauge and scalar modes is different in the different vacua, the total number of degrees of freedom is, in each case, equal to $2(N^2 - 1)$, as in the unbroken phase.

The most complicated, and most interesting, of the nontrivial vacua is the "maximal embedding" case, with $\phi_{(0)}$ given by (5.59). For this vacuum, the gauge and scalar mass spectra have additional features of note. First, this "maximal embedding" also corresponds to "maximal symmetry breaking", in the sense that in this vacuum all $N^2 - 1$ gauge degrees of freedom acquire a mass. The original $2(N^2-1)$ massive scalar modes divide equally between the scalar and gauge fields. The mass spectrum reveals an intriguing and intricate pattern, as shown in Table III. It is interesting to note that for the $SU(N)$ maximal symmetry breaking vacuum, the entire scalar mass spectrum is *almost* degenerate with the gauge mass spectrum: there is just *one* single complex component for which the masses differ!

141

gauge masses							
real fields	complex fields						
2	1	2	3	4	5	$\cdots$	N-1
6	5	8	11	14	$\cdots$	3N-4	
12	11	16	21	$\cdots$	5N-9		
20	19	26	$\cdots$	7N-16			
30	29	$\cdots$	9N-25				
$\vdots$	$\vdots$						
N(N-1)	N(N-1)-1						

scalar masses							
real fields	complex fields						
2	N	2	3	4	5	$\cdots$	N-1
6	5	8	11	14	$\cdots$	3N-4	
12	11	16	21	$\cdots$	5N-9		
20	19	26	$\cdots$	7N-16			
30	29	$\cdots$	9N-25				
$\vdots$	$\vdots$						
N(N-1)	N(N-1)-1						

Table III. $SU(N)$ mass spectrum, in units of the fundamental mass scale $\frac{v^2}{\kappa}$, for the maximal symmetry breaking vacuum, for which $\phi_{(0)}$ is given by (5.59). Notice that the gauge mass spectrum and the scalar mass spectrum are *almost* degenerate - they differ in just one complex field component.

E. Mass Matrices for Real Fields

In this Section we explain the algebraic origin of the remarkable symmetry patterns of the vacuum mass spectra for the gauge and scalar excitations in the various broken vacua of the self-dual Chern-Simons system discussed in Section V B. The most striking and significant features are that all masses are integer or half-integer multiples of the unbroken mass scale $m = v^2/\kappa$, and that in each broken vacuum the masses are *paired* so that each mass appears an even number of times. In the abelian model, discussed in Section IV E, there is just one nontrivial vacuum and in this broken vacuum there is one massive real scalar field and one massive real gauge field. The masses of these two fields are in fact equal. In the nonabelian theories the pairing of masses is more complicated. In some cases, analogous to the abelian model, this pairing arises because there is a real gauge field with mass equal to that of a real scalar field. In other cases the pairing arises because the field (either scalar or gauge) is itself a complex one (and so has two real degrees of freedom with equal mass). The former case is the more interesting and in this Section we explain how and why these gauge-scalar mass degeneracies arise in a broken vacuum. For definiteness, we present the analysis for $SU(N)$, but the generalization to other simply-laced gauge algebras is straightforward [65,66].

In a broken vacuum in which the scalar field has vacuum expectation value

$$< \phi >_{(0)} = \sum_{a=1}^{r} \phi_{(0)}^a E^a \tag{5.77}$$

the real scalar fields arise from fields ϕ with the same algebraic decomposition as $< \phi_{(0)} >$:

$$\phi = \sum_{a=1}^{r} \frac{\phi^a}{\sqrt{2}} E^a \tag{5.78}$$

143

where the ϕ^a are *real* fields. Similarly, the real gauge fields arise from diagonal gauge fields

$$A_\mu = i \sum_{a=1}^{r} A_\mu^a h_a \qquad (5.79)$$

where the hermitean Cartan subalgebra generators h_a are normalized such that

$$tr(h_a h_b) = \delta_{ab} \qquad (5.80)$$

This should be contrasted with the traces of the Chevalley basis diagonal generators H_a in (3.51) whose traces involve the Cartan matrix. For example, for the defining representation of $SU(3)$ the Chevalley basis Cartan subalgebra generators are

$$H_1 = \begin{pmatrix} 1 & 0 & 0 \\ 0 & -1 & 0 \\ 0 & 0 & 0 \end{pmatrix} \qquad H_2 = \begin{pmatrix} 0 & 0 & 0 \\ 0 & 1 & 0 \\ 0 & 0 & -1 \end{pmatrix} \qquad (5.81)$$

while the "Gell-Mann" basis Cartan subalgebra generators are

$$h_1 = \frac{1}{\sqrt{2}} \begin{pmatrix} 1 & 0 & 0 \\ 0 & -1 & 0 \\ 0 & 0 & 0 \end{pmatrix} \qquad h_2 = \frac{1}{\sqrt{6}} \begin{pmatrix} 1 & 0 & 0 \\ 0 & 1 & 0 \\ 0 & 0 & -2 \end{pmatrix} \qquad (5.82)$$

In general, the relation between the two sets of generators involves the weights of $SU(N)$:

$$h_a = \sum_{a=1}^{N-1} \omega_a^{(b)} H_b \qquad (5.83)$$

where $\vec{\omega}^{(b)}$ is the bth fundamental weight of the gauge algebra [26]. Explicitly, for $SU(N)$

$$h_a = \frac{1}{\sqrt{a(a+1)}} \sum_{b=1}^{a} b\, H_b \qquad (5.84)$$

The generators h_a satisfy the commutation relations

$$[h_a, h_b] = 0 \qquad\qquad [h_a, E_b] = \alpha_a^{(b)} E_b \qquad (5.85)$$

where $\vec{\alpha}^{(b)}$ is the bth simple root and is related to the fundamental weights $\vec{\omega}^{(b)}$ by the Cartan matrix (3.52)

$$\vec{\alpha}^{(a)} = \sum_{b=1}^{r} C_{ba}\, \vec{\omega}^{(b)} \qquad (5.86)$$

With this choice (5.79) for the gauge fields, the quadratic gauge field Lagrange density becomes

$$-\frac{\kappa}{2}\epsilon^{\mu\nu\rho} tr\left(A_\mu \partial_\nu A_\rho\right) - v^2 tr\left|\left[< \phi >_{(0)}, A_\mu\right]\right|^2$$

$$= \frac{\kappa}{2}\epsilon^{\mu\nu\rho} A_\mu^a \partial_\nu A_\rho^a - v^2 \sum_{a,b,c}\left(\phi_{(0)}^c\right)^2 \alpha_a^{(c)} \alpha_b^{(c)} A_\mu^a A^{b\mu} \qquad (5.87)$$

Therefore, recalling that the gauge masses are generated by the Chern- Simons Higgs mechanism discussed in Section IV E, the mass matrix for the real gauge fields is

$$\mathcal{M}_{ab}^{(\text{gauge})} = 2\, m \sum_{c=1}^{r}(\phi_{(0)}^c)^2\, \alpha_a^{(c)}\, \alpha_b^{(c)} \qquad (5.88)$$

where $m = \frac{v^2}{\kappa}$ is the fundamental mass scale in (5.65). For example, for the maximal embedding vacuum (5.59) in $SU(N)$ this gauge mass matrix is

$$\mathcal{M}_{ab}^{(\text{gauge})} = m \sum_{c=1}^{N-1} c(N - c)\, \alpha_a^{(c)}\, \alpha_b^{(c)} \qquad (5.89)$$

By explicit computation using the simple roots of $SU(N)$, one finds that this matrix has eigenvalues

$$2,\ 6,\ 12,\ 20,\ \ldots,\ N(N - 1) \qquad (5.90)$$

in multiples of m.

The real scalar field mass matrix is computed by taking the quadratic term (5.66) in $V(\phi + <\phi>_{(0)})$. With the field ϕ expanded as in (5.78) this quadratic potential simplifies because

$$[[<\phi>_{(0)}, <\phi>^\dagger_{(0)}], \phi] = \phi \tag{5.91}$$

and

$$[[\phi, <\phi>^\dagger_{(0)}], <\phi>_{(0)}] = [[<\phi>_{(0)}, \phi^\dagger], <\phi>_{(0)}]$$

$$= [[<\phi>_{(0)}, <\phi>^\dagger_{(0)}], \phi] \tag{5.92}$$

Therefore,

$$V_{\text{quad}}\left(\phi + <\phi>_{(0)}\right) = \frac{4v^4}{\kappa^2} tr\left(\left|[[<\phi>_{(0)}, <\phi>^\dagger_{(0)}], \phi]\right|^2\right)$$

$$= \frac{4v^4}{\kappa^2} \sum_{a,b} \phi^a \phi^b \left(\phi^a_{(0)} \phi^b_{(0)} \sum_{c=1}^{r} |\phi^c_{(0)}|^2 C_{ac} C_{bc}\right) \tag{5.93}$$

So the real scalar field mass (squared) matrix is

$$\mathcal{M}^{(\text{scalar})}_{ab} = 4\, m^2\, \phi^a_{(0)} \phi^b_{(0)} \sum_{c=1}^{r} |\phi^c_{(0)}|^2 C_{ac} C_{bc} \tag{5.94}$$

where C is the Cartan matrix (3.52). For the $SU(N)$ maximal symmetry breaking vacuum (5.59) this mass matrix is

$$\mathcal{M}^{(\text{scalar})}_{ab} = m^2 \sqrt{a\,b\,(N-a)\,(N-b)} \sum_{c=1}^{N-1} c\,(N-c)\, C_{ac} C_{bc} \tag{5.95}$$

which has eigenvalues

$$(2)^2,\ (6)^2,\ (12)^2,\ (20)^2,\ \ldots\ ,\ (N(N-1))^2 \tag{5.96}$$

in units of m^2. It is interesting to note that the eigenvalues in (5.96) are the squares of the eigenvalues (5.90) of $\mathcal{M}^{(\text{gauge})}$, even though

146

$\mathcal{M}^{\text{(scalar)}}$ is *not* the square of the matrix $\mathcal{M}^{\text{(gauge)}}$ in this basis. Nevertheless, as the real scalar masses are given by the square roots of the eigenvalues in (5.96), we see that the real scalar masses do indeed coincide with the real gauge masses. The mass spectrum of real masses may be conveniently summarized as

$$m_a = m\, a(a + 1) \qquad\qquad a = 1, 2, \ldots (N - 1) \qquad (5.97)$$

To explain the *algebraic* origin of this remarkable degeneracy between the gauge and scalar masses, and to explain the particular masses that arise, we reconsider the vacuum condition (5.54). With a factor of $|v|$ absorbed into the fields, this can be viewed as an embedding of $SU(2)$ into the original gauge algebra $\mathcal{G}$.[6] Thus, the vacuum solution $\phi_{(0)}$ may be identified with an $SU(2)$ raising operator J_+, $\phi_{(0)}^\dagger$ with J_-, and $[\phi_{(0)}, \phi_{(0)}^\dagger]$ with J_3. Then the quadratic gauge field term in (5.69) may be re-cast in terms of the adjoint action of $SU(2)$ on the gauge algebra $\mathcal{G}$:

$$m\operatorname{tr}\left(A_\mu \left(J_+ J_- + J_- J_+\right) A^\mu\right) = m\operatorname{tr}\left(A_\mu \left(\mathcal{C} - J_3^2\right) A^\mu\right) \qquad (5.98)$$

where $\mathcal{C}$ is the $SU(2)$ quadratic Casimir. But $J_3 A^\mu = 0$ since the gauge fields are restricted to the Cartan subalgebra by the ansatz (5.79). Thus the gauge masses are just given by the eigenvalues of the quadratic Casimir $\mathcal{C}$ in the adjoint action (corresponding to the particular $SU(2)$ embedding) of $SU(2)$ on the gauge algebra $\mathcal{G}$. It is a classical result of Lie algebra representation theory [70,157,32] that the adjoint action of the "principal $SU(2)$ embedding" (5.59)

---

[6]Note that this type of embedding problem also plays a significant role in the theory of instantons and of spherically symmetric magnetic monopoles [179] and the Toda molecule equations [93].

on $\mathcal{G}$ divides the $d \times d$ dimensional adjoint representation of $\mathcal{G}$ into r irreducible sub-blocks, each of dimension $(2s_a + 1)$ where the s_a are known as the *exponents* of the algebra $\mathcal{G}$. Here d is the dimension of the algebra $\mathcal{G}$. This sub-blocking fills out the entire $d \times d$ adjoint representation since the exponents have the property that

$$\sum_{a=1}^{r} (2s_a + 1) = d \tag{5.99}$$

To illustrate in detail how this decomposition works, consider for example the adjoint action of $SU(2)$ on $SU(3)$. Take a Chevalley basis for $SU(3)$

$$\{E_1, E_2, E_3, E_{-1}, E_{-2}, E_{-3}, H_1, H_2\} \tag{5.100}$$

and take

$$\phi_{(0)} = E_1 + E_2 \tag{5.101}$$

Then $\phi_{(0)}$ satisfies the vacuum condition (5.54). We therefore identify

$$J_+ = ad\left(\phi_{(0)}\right)$$

$$J_- = ad\left(\phi_{(0)}^\dagger\right)$$

$$J_3 = ad\left([\phi_{(0)}, \phi_{(0)}^\dagger]\right) \tag{5.102}$$

That is, $J_+ T \equiv [\phi_{(0)}, T]$ and $J_3 T \equiv [[\phi_{(0)}, \phi_{(0)}^\dagger], T]$ etc... Then using the Chevalley basis commutation relations we find that $J_\pm$ and J_3 act irreducibly on the $SU(3)$ generators

$$\{E_1 + E_2, H_1 + H_2, E_{-1} + E_{-2}\} \tag{5.103}$$

with

$$J_+ \begin{pmatrix} E_1 + E_2 \\ H_1 + H_2 \\ E_{-1} + E_{-2} \end{pmatrix} = \begin{pmatrix} 0 & 0 & 0 \\ -1 & 0 & 0 \\ 0 & 1 & 0 \end{pmatrix} \begin{pmatrix} E_1 + E_2 \\ H_1 + H_2 \\ E_{-1} + E_{-2} \end{pmatrix} \qquad (5.104a)$$

$$J_- \begin{pmatrix} E_1 + E_2 \\ H_1 + H_2 \\ E_{-1} + E_{-2} \end{pmatrix} = \begin{pmatrix} 0 & -1 & 0 \\ 0 & 0 & 1 \\ 0 & 0 & 0 \end{pmatrix} \begin{pmatrix} E_1 + E_2 \\ H_1 + H_2 \\ E_{-1} + E_{-2} \end{pmatrix} \qquad (5.104b)$$

$$J_3 \begin{pmatrix} E_1 + E_2 \\ H_1 + H_2 \\ E_{-1} + E_{-2} \end{pmatrix} = \begin{pmatrix} 1 & 0 & 0 \\ 0 & 0 & 0 \\ 0 & 0 & -1 \end{pmatrix} \begin{pmatrix} E_1 + E_2 \\ H_1 + H_2 \\ E_{-1} + E_{-2} \end{pmatrix} \qquad (5.104c)$$

Thus, in this irreducible sub-block,

$$\mathcal{C} \begin{pmatrix} E_1 + E_2 \\ H_1 + H_2 \\ E_{-1} + E_{-2} \end{pmatrix} \equiv \left(J_+ J_- + J_- J_+ + J_3^2 \right) \begin{pmatrix} E_1 + E_2 \\ H_1 + H_2 \\ E_{-1} + E_{-2} \end{pmatrix}$$

$$= 2 \begin{pmatrix} E_1 + E_2 \\ H_1 + H_2 \\ E_{-1} + E_{-2} \end{pmatrix} \qquad (5.105)$$

Also, $J_\pm$ and J_3 act irreducibly on the remaining 5 generators

$$\{ E_3, E_1 - E_2, H_1 - H_2, E_{-1} - E_{-2}, E_{-3} \} \qquad (5.106)$$

with

$$J_+ \begin{pmatrix} E_3 \\ E_1 - E_2 \\ H_1 - H_2 \\ E_{-1} - E_{-2} \\ E_{-3} \end{pmatrix} = \begin{pmatrix} 0 & 0 & 0 & 0 & 0 \\ -2 & 0 & 0 & 0 & 0 \\ 0 & -3 & 0 & 0 & 0 \\ 0 & 0 & 1 & 0 & 0 \\ 0 & 0 & 0 & 1 & 0 \end{pmatrix} \begin{pmatrix} E_3 \\ E_1 - E_2 \\ H_1 - H_2 \\ E_{-1} - E_{-2} \\ E_{-3} \end{pmatrix}$$

$$(5.107a)$$

$$J_- \begin{pmatrix} E_3 \\ E_1 - E_2 \\ H_1 - H_2 \\ E_{-1} - E_{-2} \\ E_{-3} \end{pmatrix} = \begin{pmatrix} 0 & -1 & 0 & 0 & 0 \\ 0 & 0 & -1 & 0 & 0 \\ 0 & 0 & 0 & 3 & 0 \\ 0 & 0 & 0 & 0 & 2 \\ 0 & 0 & 0 & 0 & 0 \end{pmatrix} \begin{pmatrix} E_3 \\ E_1 - E_2 \\ H_1 - H_2 \\ E_{-1} - E_{-2} \\ E_{-3} \end{pmatrix}$$

$$(5.107b)$$

$$J_3 \begin{pmatrix} E_3 \\ E_1 - E_2 \\ H_1 - H_2 \\ E_{-1} - E_{-2} \\ E_{-3} \end{pmatrix} = \begin{pmatrix} 2 & 0 & 0 & 0 & 0 \\ 0 & 1 & 0 & 0 & 0 \\ 0 & 0 & 0 & 0 & 0 \\ 0 & 0 & 0 & -1 & 0 \\ 0 & 0 & 0 & 0 & -2 \end{pmatrix} \begin{pmatrix} E_3 \\ E_1 - E_2 \\ H_1 - H_2 \\ E_{-1} - E_{-2} \\ E_{-3} \end{pmatrix}$$

$$(5.107c)$$

Thus in this irreducible sub-block,

$$\mathcal{C} \begin{pmatrix} E_3 \\ E_1 - E_2 \\ H_1 - H_2 \\ E_{-1} - E_{-2} \\ E_{-3} \end{pmatrix} = 6 \begin{pmatrix} E_3 \\ E_1 - E_2 \\ H_1 - H_2 \\ E_{-1} - E_{-2} \\ E_{-3} \end{pmatrix}$$

$$(5.108)$$

150

So, the adjoint action of $SU(2)$ on $SU(3)$ decomposes into one spin 1 representation of dimension $3 = 2(1) + 1$ and one spin 2 representation of dimension $5 = 2(2) + 1$. This corresponds to the exponents $s_1 = 1$ and $s_2 = 2$ of $SU(3)$.

The elements of each irreducible sub-block are arranged according to their corresponding *principal grading* which is their J_3 eigenvalue. Restricting to the Cartan subalgebra (as is achieved by the ansatz (5.79) for the gauge fields) selects the $j_3 = 0$ element from each sub-block, and in the ath irreducible sub-block the quadratic Casimir $\mathcal{C}$ has eigenvalue

$$\mathcal{C} = s_a(s_a + 1) \tag{5.109}$$

See, for example, (5.105) and (5.108). Thus, from the discussion immediately following equation (5.98) we deduce that the gauge masses should be m times these eigenvalues $s_a(s_a + 1)$ of $\mathcal{C}$.

Table IV displays the eigenvalues of the gauge mass matrix (5.88) for the simply-laced Lie algebras, obtained by explicit computation using the simple roots of these algebras. The exponents s_a for the classical simply-laced Lie algebras are listed in Table V. It is straightforward to verify that the mass spectra in Table IV for the eigenvalues of the gauge mass matrix (5.88) do indeed coincide with the general mass formula

$$m_a = m\, s_a(s_a + 1) \qquad\qquad a = 1, \ldots r \tag{5.110}$$

where the s_a are the *exponents* of $\mathcal{G}$. (It is interesting to note that the mass spectrum of the affine Toda theory is also given in terms of the exponents of the gauge algebra [88,89].)

Algebra	Masses							
A_r	2	6	12	20	30	$\cdots$	$r(r-1)$	$r(r+1)$
D_r	2	12	30	56	90	$\ldots$	$(2r-3)(2r-2)$	$r(r-1)$
E_6	2	20	30	56	72	132		
E_7	2	30	56	90	132	182	306	
E_8	2	56	132	182	306	380	552	870

Table IV. The gauge masses, in units of m, for the principal embedding vacuum (5.59), obtained as square roots of the eigenvalues of the mass matrix in (5.88). Comparing with Table V we see that the gauge masses are related to the exponents of the gauge algebra by the relation (5.110).

Algebra	Rank	Dim.	Exponents							
A_r	r	$r(r+2)$	1	2	3	$\cdots$	$r-1$	r		
D_r	r	$r(2r-1)$	1	3	5	$\ldots$	$2r-3$	$r-1$		
E_6	6	78	1	4	5	7	8	11		
E_7	7	133	1	5	7	9	11	13	17	
E_8	8	248	1	7	11	13	17	19	23	29

Table V. The ranks, dimensions and exponents of the simply-laced classical Lie algebras. Note that the sum of the exponents equals the number of positive roots, which is one half (dimension – rank).

To see that the real scalar masses are also given by the general mass formula (5.110), we note that the quadratic part (5.66) of the shifted scalar potential can be written as

$$4m^2 \text{tr}\left(\phi^\dagger \left(J_+ J_-\right)^2 \phi\right) = m^2 \text{tr}\left(\phi^\dagger \left(\mathcal{C} - J_3^2 + J_3\right)^2 \phi\right) \qquad (5.111)$$

But $J_3 \phi = 1\phi$ since ϕ is expanded in terms of the simple root step operators (and hence has principal grading 1). Thus the eigenvalues of the scalar $mass^2$ matrix are the squares of the eigenvalues of $\mathcal{C}$, and we find a scalar mass spectrum identical with the gauge mass spectrum in (5.110).

For any vacuum $\phi_{(0)}$ *other* than the principal embedding one (5.59), the gauge and scalar masses may be found as follows. If the vacuum solution $\phi_{(0)}$ corresponds to n deletions of dots from the original Dynkin diagram (as described before) then n complex scalar fields remain massive with mass m corresponding to the scalar mass in the unbroken vacuum. The remaining $(r-n)$ real scalar masses are obtained from formula (5.110) using the exponents for each of the Dynkin sub-diagrams. This also yields the $(r-n)$ real gauge masses. Thus in any vacuum, the masses are always paired, either because they correspond to a complex scalar field (of which the extreme case is the unbroken vacuum) or because the real scalar and gauge masses coincide through formula (5.110) (of which the principal embedding vacuum (5.59) is the extreme case).

VI. QUANTUM ASPECTS

In this Chapter we conclude our discussion of the self-dual Chern-Simons systems by addressing various *quantum* aspects of the models. The previous Chapters have concentrated on classical aspects, although particular emphasis has been given to that classical data (such as static solutions, and vacuum spectra) which is usually of the greatest significance for quantization [122,211]. In $1 + 1$ dimensional field theories, the equivalence between direct quantization and the quantization *about* classical soliton solutions has been explored in great detail, leading to major advances and applications in both physics and mathematics [82,211,212]. This is true both of nonrelativistic models such as the $1 + 1$ dimensional nonlinear Schrödinger equation, and of relativistic models such as the $1 + 1$ dimensional Sine-Gordon model. In such systems, *integrability* in $1 + 1$ dimensions plays a crucial role. In the $2 + 1$ dimensional self-dual Chern-Simons systems, there are no clear indications of complete $2 + 1$ dimensional integrability (despite the close links between the *static* self-dual solutions and two dimensional Euclidean integrable models, as discussed in Chapters 2 and 3), and so it has not been possible to carry through in such explicit detail the quantization of the self-dual Chern-Simons systems. Nevertheless, a great deal has been achieved, and this Chapter contains a (personalized) summary of the current state of affairs. For the nonrelativistic self-dual Chern-Simons models, quantization has led to beautiful advances in the understanding of anyon physics, of Aharonov-Bohm scattering, of scale invariance in planar systems, and of delta-function interactions in planar models. In the relativistic self-dual Chern-Simons systems, the most definitive and comprehensive results thus far are concerned with the fate of the extended supersymmetry properties of the self-dual theories at the quantum level.

A. Nonrelativistic Matter–Chern–Simons Field Theory

It is well known that nonrelativistic quantum field theory provides a second-quantized formulation of N-particle nonrelativistic quantum mechanics. This correspondence permits an efficient approach to the calculation of certain N-body processes in quantum mechanics, using the powerful machinery of quantum field theory. In this Section we show that this approach may be extended to incorporate the Chern-Simons fields present in the nonrelativistic self-dual Chern-Simons systems [126]. Since the effect of a Chern-Simons gauge field is to constrain a 'magnetic' flux to the matter density, it is not so surprising that the resulting quantum mechanical model is that of the multi-anyon system. Anyons are nonrelativistic[7] point particles in two spatial dimensions that carry both electric charge and magnetic charge - crudely speaking, they may be considered as composites of point charged particles with magnetic flux lines 'attached' [247,175]. While this expectation is very natural, the explicit implementation of the correspondence between the nonrelativistic Chern-Simons-matter quantum field theory and nonrelativistic multi-anyon quantum mechanics requires some care and delicacy, as discussed below. This subject has been discussed in various related contexts in many places, and I recommend that the interested reader consult complementary references in order to appreciate the many different applications and interpretations. My discussion follows quite closely that of Lerda [175] and is focussed

------

[7]The term 'anyon' was originally applied to nonrelativistic systems, although it has since been extrapolated to relativistic planar systems with fractional spin and statistics .

155

primarily on illustrating the important role of *self-duality* in the quantum theory. For further details more directly related to anyon physics and its applications the reader should consult, for example, [247,175,234,85,223,118,194].

To second-quantize a classical nonrelativistic field theory involving the complex scalar field $\psi(\vec{x}, t)$, and its conjugate $\psi^\dagger(\vec{x}, t)$, we promote these classical fields to *operator-valued* fields satisfying the fundamental equal-time commutation relations:

$$[\psi(\vec{x}, t), \psi(\vec{y}, t)] = 0$$

$$[\psi^\dagger(\vec{x}, t), \psi^\dagger(\vec{y}, t)] = 0$$

$$[\psi(\vec{x}, t), \psi^\dagger(\vec{y}, t)] = \delta(\vec{x} - \vec{y}) \tag{6.1}$$

These field operators evolve in time according to the Heisenberg equation of motion

$$i\hbar \frac{\partial}{\partial t} \psi(\vec{x}, t) = [\psi(\vec{x}, t), H] \tag{6.2}$$

where $H = \int d^2 x \mathcal{H}$ is the Hamiltonian operator, with Hamiltonian density $\mathcal{H}$. In general H will involve bilinear (and higher) terms in the field operators and so we must prescribe an operator ordering prescription. We choose the conventional "normal-ordering" prescription in which all the $\psi^\dagger$ operators are placed to the left of the ψ operators. This is motivated by the standard Fock space construction based upon a vacuum state $|\Omega >$ which is assumed to be annihilated by ψ:

$$\psi(\vec{x})|\Omega > = 0$$
$$< \Omega|\psi^\dagger(\vec{x}) = 0 \tag{6.3}$$

In a nonrelativistic system there is no particle creation (or annihilation) and so the number density operator

$$\rho(\vec{x}, t) = \psi^\dagger(\vec{x}, t)\psi(\vec{x}, t) \tag{6.4}$$

satisfies a continuity equation which implies the conservation of the number operator

$$\hat{N} = \int d^2x\, \rho \tag{6.5}$$

Thus $\hat{N}$ and H commute, and may therefore be simultaneously diagonalized, with a common set of eigenstates $|E, N>$:

$$
\begin{aligned}
H|E, N> &= E|E, N> \\
\hat{N}|E, N> &= N|E, N>
\end{aligned}
\tag{6.6}
$$

The number operator annihilates the vacuum state $|\Omega>$, and we can redefine the minimum of H so that the Hamiltonian operator does also:

$$\hat{N}|\Omega>= 0 \qquad\qquad H|\Omega>= 0 \tag{6.7}$$

The Fock space states are generated from the vacuum state $|\Omega>$ by the action of the creation operators $\psi^\dagger$:

$$|\vec{x}_1, \vec{x}_2, \ldots, \vec{x}_N> = \psi^\dagger(\vec{x}_N)\ldots \psi^\dagger(\vec{x}_2)\psi^\dagger(\vec{x}_1)|\Omega> \tag{6.8}$$

Since $\hat{N}$ is conserved, the Fock space may be decomposed into N-particle sectors by defining the corresponding N-body quantum mechanical wavefunction

$$\Psi_E(\vec{x}_1, \vec{x}_2, \ldots, \vec{x}_N) \equiv< \vec{x}_1, \vec{x}_2, \ldots, \vec{x}_N|E, N> \tag{6.9}$$

By construction, this N-particle wavefunction is totally symmetric with respect to interchange of particle labels. This is a direct consequence of the fundamental commutation relations (6.1). (For multifermion systems we simply replace these commutation relations with corresponding anti-commutation relations.)

157

As a consequence of (6.6) and (6.7) we see that

$$E\Psi_E(\vec{x}_1, \vec{x}_2, \ldots, \vec{x}_N) = <\Omega|\,[\psi(\vec{x}_1)\psi(\vec{x}_2)\ldots\psi(\vec{x}_N), H]\,|E, N>$$

(6.10)

The commutator appearing on the right-hand-side of (6.10) may be evaluated once we know the form of the Hamiltonian operator H, thereby yielding the N-particle Schrödinger equation satisfied by the N-particle wavefunction $\Psi_E(\vec{x}_1, \vec{x}_2, \ldots, \vec{x}_N)$. For example, for a free Hamiltonian

$$H = \frac{1}{2m} \int d^2x\, (\vec{\nabla}\psi)^\dagger \vec{\nabla}\psi$$

(6.11)

we obtain the free N-body Schrödinger equation:

$$-\frac{1}{2m}\left(\nabla_1^2 + \nabla_2^2 + \ldots + \nabla_N^2\right)\Psi_E(\vec{x}_1, \vec{x}_2, \ldots, \vec{x}_N)$$

$$= E\Psi_E(\vec{x}_1, \vec{x}_2, \ldots, \vec{x}_N)$$

(6.12)

For the purposes of the nonrelativistic self-dual Chern-Simons systems we wish to generalize this (standard) formalism in two important ways.

First, we include a quartic $|\psi|^4$ potential term in the Hamiltonian operator H:

$$H = \frac{1}{2m} \int d^2x\, (\vec{\nabla}\psi)^\dagger \vec{\nabla}\psi - \frac{g}{2} \int d^2x\, :(\psi^\dagger\psi)^2:$$

(6.13)

where the : : symbols denote normal ordering. Such a quartic term appears in the *classical* nonrelativistic self-dual Chern-Simons Lagrange density (2.33). In the corresponding N-body nonrelativistic quantum mechanical system, we find from (6.10) that this potential corresponds to delta-function interactions between the nonrelativistic point particles. The *sign* of the coupling g determines whether these delta-function interactions are repulsive or attractive. For example, in the two-particle sector, the Schrödinger equation corresponding to the Hamiltonian (6.13) becomes

$$-\frac{1}{2m}\left(\nabla_1^2 + \nabla_2^2\right)\Psi_E(\vec{x}_1, \vec{x}_2) - g\delta(\vec{x}_1 - \vec{x}_2)\Psi_E(\vec{x}_1, \vec{x}_2) = E\Psi_E(\vec{x}_1, \vec{x}_2)$$

(6.14)

It is straightforward to generalize this to the N-particle case:

$$-\frac{1}{2m}\left(\nabla_1^2 + \nabla_2^2 + \ldots + \nabla_N^2\right)\Psi_E(\vec{x}_1, \vec{x}_2, \ldots, \vec{x}_N)$$

$$- g\sum_{I<J}\delta(\vec{x}_I - \vec{x}_J)\Psi_E(\vec{x}_1, \vec{x}_2, \ldots, \vec{x}_N)$$

$$= E\Psi_E(\vec{x}_1, \vec{x}_2, \ldots, \vec{x}_N)$$

(6.15)

It is important to stress that this formal correspondence between quantum mechanical delta-function interactions and a quartic $|\psi|^4$ potential term in the second-quantized Hamiltonian operator H is not dependent on the spatial dimension - it is direct consequence of the relations (6.1). Indeed, the first serious study of the implications of a nonrelativistic $|\psi|^4$ potential in quantum field theory arose in the work of Bég and Furlong [17] who were investigating the question of the triviality of relativistic $\lambda\phi^4$ theory (in 4 dimensional spacetime) by a consideration of its nonrelativistic limit. However, we shall see in the next Section that for planar (*i.e.* two space dimensional) theories, the treatment of delta-function interactions possesses important and interesting subtleties which ultimately may be traced to the fact that ∇^2 and $\delta^{(2)}(\vec{x})$ have the same scaling dimension in two dimensional space. In fact, these subtleties have important *positive* consequences which we may exploit in the quantal analysis of nonrelativistic self-dual Chern-Simons theories.

The second major extension of the standard correspondence between nonrelativistic quantum field theory and N-body quantum mechanics is to couple the matter fields to a Chern-Simons gauge field A_μ, as in the classical self-dual Chern-Simons systems discussed in Chapters 2 and 3. Thus, we wish to investigate the quantum behavior of the Hamiltonian operator (see Section II B for the corresponding classical discussion)

$$H = \frac{1}{2m} \int d^2x \, (\vec{D}\psi)^\dagger \vec{D}\psi - \frac{g}{2} \int d^2x \, : (\psi^\dagger \psi)^2 : \tag{6.16}$$

where the ordinary gradient $\vec{\nabla}$ has now been replaced by its gauge covariant form $\vec{D}$

$$\vec{D}\psi = \vec{\nabla}\psi + i\vec{A}\psi \tag{6.17}$$

Furthermore, the vector potential $\vec{A}$ is constrained by the Chern-Simons Gauss law constraint (2.8a) to be proportional to the number density ρ:

$$\epsilon^{ij}\partial_i A_j = \frac{1}{\kappa}\rho \tag{6.18}$$

Thus, the vector potential may be expressed as (see Equation (2.17))

$$A_i(\vec{x}) = \frac{1}{\kappa}\epsilon^{ij}\frac{\partial}{\partial x^j}\int d^2y \, G(\vec{x} - \vec{y})\rho(\vec{y}) \tag{6.19}$$

where $G(\vec{x} - \vec{y}) = \frac{1}{2\pi}ln|\vec{x} - \vec{y}|$ is the planar Green's function.

The relation (6.19) suggests possible operator ordering difficulties in the Hamiltonian operator (6.16). However, the commutation relations (6.1) imply that

$$[A_i(\vec{x}), \psi(\vec{y})] = -\frac{1}{\kappa}\epsilon^{ij}\frac{\partial}{\partial x^j}G(\vec{x} - \vec{y})\psi(\vec{y}) \tag{6.20}$$

which vanishes at coincident points $\vec{x} = \vec{y}$ if carefully regulated so that $\epsilon^{ij}\partial_j G$ vanishes at the origin (as is the case for a regularization

160

that preserves the antisymmetry under spatial reflection - for an explicit regularization see [175]). Thus

$$[A_i(\vec{x}), \psi(\vec{x})] = 0 \qquad (6.21)$$

and so there are, in fact, no ordering problems in the kinetic part of the Hamiltonian operator H in (6.16), provided of course that we consistently maintain the same regularization procedure for manipulating the Green's function and its derivatives. Thus, H may be used to define the time evolution in the nonrelativistic quantum field theory, and the corresponding N-body Schrödinger equation for the first-quantized theory may be derived from (6.10) as before. Our task now is to determine the first-quantized manifestation of the Chern-Simons coupling in (6.16).

The derivation of the N-body Schrödinger equation arising from the Hamiltonian operator H in (6.16) is notationally cumbersome, so we present here the 2-body case, and leave the generalization to the N-body case as an exercise for the reader. Since only 2-body interactions are involved, the generalization is in fact straightforward. We first need the commutator $[\psi(\vec{x}_1), H]$ where H is the Hamiltonian H in (6.16). From the commutation relations (6.1) and (6.20) we find

$$[\psi(\vec{x}_1), H] = -\frac{1}{2m} \nabla_1^2 \psi(\vec{x}_1) + \frac{1}{2m} \vec{A}(\vec{x}_1)^2 \psi(\vec{x}_1) - g \psi^\dagger(\vec{x}_1) \psi^2(\vec{x}_1)$$

$$- \frac{i}{2m} \left[\vec{A}(\vec{x}_1) \cdot \vec{\nabla}_1 \psi(\vec{x}_1) + \vec{\nabla}_1 \cdot \left(\vec{A}(\vec{x}_1) \psi(\vec{x}_1) \right) \right]$$

$$- \frac{i}{2m\kappa} \int d^2 y \, \epsilon^{ij} \frac{\partial}{\partial y^j} G(\vec{y} - \vec{x}_1) \psi(\vec{x}_1) \left[\psi^\dagger(\vec{y}) \frac{\partial}{\partial y^i} \psi(\vec{y}) - \frac{\partial}{\partial y^i} \psi^\dagger(\vec{y}) \psi(\vec{y}) \right]$$

$$- \frac{1}{2m\kappa} \int d^2 y \, \epsilon^{ij} \left[\frac{\partial}{\partial y^j} G(\vec{y} - \vec{x}_1) \psi(\vec{x}_1) A^i(\vec{y}) \rho(\vec{y}) \right]$$

$$-\frac{1}{2m\kappa}\int d^2y\,\epsilon^{ij}\left[A^i(\vec{y})\frac{\partial}{\partial y^j}G(\vec{y}-\vec{x}_1)\psi(\vec{x}_1)\rho(\vec{y})\right] \qquad (6.22)$$

Note that the g dependent term arises from the quartic potential in H, while the κ dependent terms arise due to the commutation relation (6.20). This expression may be simplified by using the relation between the current density $\vec{J}$ and A^0

$$A_0(\vec{x}) = -\frac{1}{\kappa}\int d^2y\,G(\vec{x}-\vec{y})\epsilon^{ij}\partial_i J_j(\vec{y}) \qquad (6.23)$$

leading to

$$[\psi(\vec{x}_1)\ ,\ H] = -\frac{1}{2m}\nabla_1^2\psi(\vec{x}_1) - A^0(\vec{x}_1)\psi(\vec{x}_1) - g\psi^\dagger(\vec{x}_1)\psi^2(\vec{x}_1)$$

$$+\frac{1}{2m\kappa^2}\int d^2y\,\vec{\nabla}G(\vec{y}-\vec{x}_1)\cdot\vec{\nabla}G(\vec{y}-\vec{x}_1)\rho(\vec{y})\psi(\vec{x}_1) \qquad (6.24)$$

The last term in this expression is a quantum re-ordering term. Now, both A^0 and ρ annihilate the vacuum from the right, so that

$$<\Omega|[\psi(\vec{x}_1),H] = -\frac{1}{2m}\nabla_1^2<\Omega|\psi(\vec{x}_1) \qquad (6.25)$$

Having computed the commutator of a single field operator $\psi(\vec{x}_1)$ with H we can now use the results (6.24) and (6.25) to compute the 2-body commutator appearing in the Schrödinger equation (6.10) as

$$<\Omega|[\psi(\vec{x}_1)\psi(\vec{x}_2),H]|E,2>=$$

$$-\frac{1}{2m}\sum_{I=1}^{2}\left(\vec{\nabla}_I + i\vec{A}_I(\vec{x}_1,\vec{x}_2)\right)^2<\Omega|\psi(\vec{x}_1)\psi(\vec{x}_2)|E,2>$$

$$-g\delta(\vec{x}_1-\vec{x}_2)<\Omega|\psi(\vec{x}_1)\psi(\vec{x}_2)|E,2> \qquad (6.26)$$

where the (nonlocal) 'vector potential' $\vec{A}_I(\vec{x}_1,\vec{x}_2)$ is given by

$$A_I^i(\vec{x}_1,\vec{x}_2) = -\frac{1}{\kappa}\epsilon^{ij}\frac{\partial}{\partial x_I^j}\left[\sum_{J\neq I}G(\vec{x}_I-\vec{x}_J)\right]$$

162

$$= -\frac{1}{2\pi\kappa}\epsilon^{ij}\sum_{J\neq I}\frac{x_I^j - x_J^j}{|\vec{x}_I - \vec{x}_J|^2} \qquad (6.27)$$

Thus (6.26) leads to the 2-body Schrödinger equation

$$\left(-\frac{1}{2m}\sum_{I=1}^{2}\left(\vec{\nabla}_I + i\vec{A}_I(\vec{x}_1, \vec{x}_2)\right)^2 - g\delta(\vec{x}_1 - \vec{x}_2)\right)\Psi_E(\vec{x}_1, \vec{x}_2)$$

$$= E\Psi_E(\vec{x}_1, \vec{x}_2) \qquad (6.28)$$

with $\vec{A}_I(\vec{x}_1, \vec{x}_2)$ given in (6.27) - this is precisely the 2-anyon Schrödinger equation. The generalization to the N-particle case is straightforward, and one sees that the Schrödinger equation coming from the nonrelativistic quantum field theory with Hamiltonian H is precisely equivalent to the N-particle anyon system [126,175], with an additional δ-function interaction.

The first-quantized Hamiltonian h corresponding to H is

$$h = -\frac{1}{2m}\sum_{I=1}^{2}\left(\vec{\nabla}_I + i\vec{A}_I(\vec{x}_1, \vec{x}_2)\right)^2 - g\delta(\vec{x}_1 - \vec{x}_2)$$

$$= -\frac{1}{2m}\sum_{I=1}^{2}D_+^I D_-^I + \left(\frac{1}{m\kappa} - g\right)\delta(\vec{x}_1 - \vec{x}_2) \qquad (6.29)$$

where $D_\pm^I$ are defined as

$$D_\pm^I \equiv D_1^I \pm iD_2^I \qquad (6.30)$$

with

$$\vec{D}^I \equiv \vec{\nabla}_I + i\vec{A}_I \qquad (6.31)$$

The extra delta function interaction term in the first-quantized Hamiltonian (6.29) arises from the commutator term $[D_+, D_-]$, using the relation

$$\epsilon^{ij}\partial_i\partial_j\theta = 2\pi\delta(\vec{x})\tag{6.32}$$

For the N-particle case, the first-quantized Hamiltonian h corresponding to H is

$$h = -\frac{1}{2m}\sum_{I=1}^{N}D_+^I D_-^I + \left(\frac{1}{m\kappa} - g\right)\sum_{I<J}^{N}\delta(\vec{x}_I - \vec{x}_J)\tag{6.33}$$

where $\vec{A}_I$ is now the N-particle generalization of (6.27). Thus, if we choose the quartic coupling g to take the self-dual critical value (2.32)

$$g = \frac{1}{m\kappa}\tag{6.34}$$

then the delta-function interactions disappear from the normal-ordered first-quantized Hamiltonian in (6.33). It is precisely at this value of the coupling g that the classical nonrelativistic Chern-Simons field theory has static self-dual solutions. In the quantum theory, this self-duality manifests itself as a cancellation between the magnetic interaction term coming from interactions between the magnetic fluxes of the anyons and the magnetic interaction term coming from

$$-\frac{g}{2} : (\psi^\dagger(\vec{x})\psi(\vec{x}))^2 := -\frac{g\kappa}{2} : B(\vec{x})\rho(\vec{x}) :\tag{6.35}$$

where we have used the Chern-Simons Gauss law constraint (2.8a) to rewrite the $|\psi|^4$ potential term as a Pauli magnetic interaction term. At the self-dual point these two magnetic interactions balance and cancel one another.

The self-dual first-quantized Hamiltonian in (6.33) has been studied as an exactly soluble model for the multi-anyon problem [99], and applied to the problem of the quantum Hall effect [78,79,232,233]. In the next two Sections we shall consider in some detail two other problems where this exact cancellation between the delta-function interactions plays a vital role in the successful quantization of non-relativistic planar models.

This abelian analysis may also be extended to the case of a nonabelian Chern-Simons theory. In this case the nonrelativistic self-dual Chern-Simons theory described in Chapter 3 becomes a quantum field theory which provides a second-quantized description of the N-anyon problem with nonabelian fractional statistics. The N-particle Schrödinger equation for such a system with nonabelian fractional statistics may be motivated by representations of the braid group [239], but the most direct approach (in a computational sense) is through the corresponding Chern-Simons field theory [12]. The nonabelian group structure introduces additional subtleties, primarily related to the representation of the vector potential in terms of the matter fields. There are different ways to achieve this representation, corresponding to different choices of gauge. One approach is to use a non-hermitean gauge which exploits the natural holomorphic structure of the Chern-Simons theory [76,25]. For $SU(N)$ in particular, this leads to an elegant description [173,174] in terms of generalized coherent states and Grassmannians [202]. This approach is largely motivated by the relation of Chern-Simons theories to conformal field theories [249,76,25]. Another approach [12] uses a real gauge to show that the nonabelian quantum theory is consistent, and that it corresponds to a second-quantized description of a system of point particles carrying nonabelian charges and non-abelian fractional statistics. This somewhat unusual point particle model is, in turn, related to the "Wong equations" which describe a nonabelian generalization of point particle magneto-electrodynamics [250]. For a systematic study of the quantization of these nonabelian systems the reader is referred to [12,173,174,154].

B. Scale Invariance in Quantized Planar Models

The relationship between the nonrelativistic self-dual Chern-Simons quantum field theory and anyonic quantum mechanics discussed in the previous Section has the consequence that one may use the powerful methods of quantum field theory as a tool to probe the multi-anyon problem. But before this is attempted, we must first address an unexpected feature of planar quantum mechanical systems with delta-function interactions. Bergman [20] has shown that the classical scale invariance of such a planar system with delta-function interactions is broken when the system is quantized. The quantization of the model requires regularization and renormalization, and as a consequence a scale must be introduced which violates the classical scale invariance. This is a beautiful example of an application of field theoretic techniques to a quantum mechanical problem, and provides the first clear evidence of the 'anomaly' phenomenon (the quantum breaking of a classical symmetry) in a quantum mechanical system. I stress that Bergman's result [20] concerns planar nonrelativistic systems with δ-function interactions, with no coupling to Chern-Simons fields. However, later in this Section we shall incorporate also Chern-Simons fields and find that this anomaly phenomenon takes on an even more interesting light.

We saw in the previous Section that the appearance of a quartic $|\psi|^4$ potential in the second-quantized Hamiltonian leads to delta-function interactions between point particles in the corresponding first-quantized theory. The work of Bég and Furlong [17] showed that in such models there appear ultraviolet divergences which require regularization and renormalization. The physical source of these divergences lies in the *zero-range* nature of the delta-function interactions. The quantum fieldtheoretic formalism (albeit nonrelativistic rather than the more familiarrelativistic one) is much better suited to the handling of such regularization issues than is conventional N-body quantum mechanics. As we discuss below, it is possible to perform an infinite renormalization, yielding finite quantities (such as scattering amplitudes) parametrized by a finite renormalized coupling strength of the delta-function interaction. Alternatively, but equivalently, the delta-function interaction may be viewed as a self-adjoint extension of a formally hermitean noninteracting Hamiltonian which is defined on a space with one (or more) point(s) removed. The renormalized coupling strength may then be identified with the parameter of the self-adjoint extension.

The situation is even more interesting in *planar* (*i.e.* two space dimensional) theories because the classical system of point particles in theplane interacting via delta-function interactions is scale invariant. This fact may be understood at the first-quantized level by observing that in the Schrödinger equation the kinetic term ∇^2 has the same scaling dimension as the delta-function potential $\delta^{(2)}(\vec{x})$. In field theoretic language, consider the Lagrange density

$$\mathcal{L} = i\psi^\dagger \partial_t \psi - \frac{1}{2}|\vec{\nabla}\psi|^2 + \frac{g}{2}|\psi|^4 \qquad (6.36)$$

Note that the nonrelativistic mass may be factored out as an overall (irrelevant) scale [20]. In these units, the gradient $\vec{\nabla}$ has engineering dimension 1, the time derivative ∂_t has dimension 2, and the quartic coupling g is dimensionless. Thus, the Lagrange density (6.36) contains no dimensionful couplings, and so is explicitly scale invariant. Consider a nonrelativistic scale transformation:

$$\vec{x} \to e^\lambda \vec{x}$$

$$t \to e^{2\lambda}t \qquad (6.37)$$

Under an infinitesimal such rescaling the scalar field ψ changes by

167

$$\delta\psi = \lambda \left(1 + \vec{x} \cdot \vec{\nabla} + 2t\partial_t\right)\psi \tag{6.38}$$

The Lagrange density $\mathcal{L}$ therefore changes by

$$\delta\mathcal{L} = \lambda \left(4 + \vec{x} \cdot \vec{\nabla} + 2t\partial_t\right)\mathcal{L} \tag{6.39}$$

This variation of $\mathcal{L}$ may be re-expressed as a total derivative

$$\delta\mathcal{L} = \lambda \left(\vec{\nabla} \cdot (\vec{x}\mathcal{L}) + 2\partial_t (t\mathcal{L})\right) \tag{6.40}$$

Thus, the classical action

$$S = \int d^2x \, dt \, \mathcal{L} \tag{6.41}$$

remains invariant (with suitable boundary conditions on the fields). The Noether charge and current density corresponding to this classical symmetry satisfy

$$\partial_t \rho + \vec{\nabla} \cdot \vec{j} = 2\Theta^{00} - \sum_{i=1}^{2} \Theta^{ii} \tag{6.42}$$

where $\Theta^{\mu\nu}$ is the nonrelativistic energy-momentum tensor. Thus, scale invariance at the classical level implies that the nonrelativistic energy momentum tensor satisfies (see also Section II D)

$$2\Theta^{00} = \sum_{i=1}^{2} \Theta^{ii} \tag{6.43}$$

This should be contrasted with the (perhaps more familiar) *relativistic* tracelessness condition

$$\Theta^{\mu}{}_{\mu} = 0 \tag{6.44}$$

satisfied by the energy-momentum tensor in a scale invariant relativistic field theory.

In the *quantum* theory, this classical condition (6.43) may be translated into a Ward identity to be satisfied by the 2n-point functions of the perturbative quantum field theory. For example, scale invariance implies a scaling relation for the 4-point function which may be checked perturbatively. Bergman [20] has shown that this scaling relation is violated by the renormalized 4-point function, and that the violation may be identified with an anomaly equation

$$2\Theta^{00} - \sum_{i=1}^{2} \Theta^{ii} = \frac{g^2}{2\pi} : (\psi^\dagger\psi)^2 : \qquad (6.45)$$

This result may seem potentially troublesome for the quantization of the nonrelativistic self-dual Chern-Simons models, since they involve a quartic $|\psi|^4$ potential which leads to delta-function interactions and which therefore breaks the scale invariance of the classical model. However, it is precisely at this point that the self-duality of the model comes into play.

We now extend the discussion of the quantum fate of classical scale invariance to include the coupling of the nonrelativistic matter fields ψ to a Chern-Simons gauge field A_μ. We therefore generalize the Lagrange density (6.36) to

$$\mathcal{L} = i\psi^\dagger D_0\psi - \frac{1}{2}|\vec{D}\psi|^2 + \frac{\kappa}{2}\epsilon^{\mu\nu\rho}A_\mu\partial_\nu A_\rho + \frac{g}{2}|\psi|^4 \qquad (6.46)$$

The Chern-Simons coupling parameter (like the quartic coupling g) is dimensionless, so the Lagrange density is still scale invariant. It is natural to ask then whether or not this scale invariance also suffers from the scale anomaly phenomenon. This issue may be explored in a number of ways. Lozano [184] has studied this question in terms of the 1-loop effective potential, while Bergman and Lozano have explored the scale invariance of this system in the context of Aharonov-Bohm scattering amplitudes [21]. In each case the result is that scale invariance is broken at the quantum level, *unless* the quartic coupling g is chosen to take its self-dual value

$$g = \pm\frac{1}{\kappa} \qquad (6.47)$$

as in (2.32). Lozano [184] computed the ground state energy for a gas of nonrelativistic bosons minimally coupled to a Chern-Simons gauge field, together with self-interactions via delta-function potentials (see also a related analysis in [14]). There are 1-loop contributions to the energy, but this additional contribution vanishes if g takes its self-dual value. As these 1-loop terms involve a renormalization scale, scale invariance is broken unless the couplings are chosen to make these terms cancel. This result may be restated [108] in terms of 1-loop contributions to the energy density, pressure and particle density in such a way that

$$2\Theta^{00} - \sum_{i=1}^{2} \Theta^{ii} = \left(g^2 - \frac{1}{\kappa^2}\right) \frac{\rho^2}{2\pi} \tag{6.48}$$

Thus, these one-loop corrections violate scale invariance except at the self-dual point where $g = \pm\frac{1}{\kappa}$. The importance of the self-dual point has also been identified in one-loop investigations of the effective potential (for both abelian and nonabelian theories) of the nonrelativistic Chern-Simons matter system [27,28]. The one-loop correction to the effective potential vanishes at the self-dual point, and away from the self-dual point a scale dependence is introduced. Higher-loop computations of related aspects of nonrelativistic matter-Chern-Simons quantum field theory, with applications to the many-anyon problem, may be found in [87,236,77].

C. Aharonov–Bohm Scattering and Chern–Simons Theory

Aharonov-Bohm scattering, *viz.* the scattering of charged particles from a magnetic flux tube, is essentially the same problem as the scattering of two anyons, when transformed to the center-of-mass frame. Therefore, the nonrelativistic Chern-Simons field theoretic description of multi-anyon quantum mechanics discussed in

Section VIA may be used also to study Aharonov-Bohm scattering. Aharonov-Bohm scattering presents an unusual problem, in the sense that the *exact* solution (for example, for the 2-body scattering amplitude) has been known for many years [2,215], and yet the *perturbative* analysis of the scattering problem has only very recently [21] been resolved. Earlier attempts at a direct quantum mechanical perturbative treatment had failed for a number of reasons. The naive Born approximation misses the s-wave first order contribution to the scattering amplitude, and the second order term in the Born series is divergent [3,195]. This failure may be understood physically in terms of the observation that the scattering from a finite-radius flux tube is ill-defined in the zero radius (*i.e.* ideal flux tube) limit [3,107].

The resolution [21] of this puzzle came from a nonrelativistic quantum field theoretic analysis using Chern-Simons field theory. The proposal for using such field theoretic techniques for Aharonov-Bohm scattering was first made by Hagen [105] in his analysis of Galilean invariant quantum field theory. Bergman and Lozano generalized Hagen's model to incorporate the $|\psi|^4$ contact interaction, which corresponds to the inclusion of delta-function potential interactions at the quantum mechanical level. Formally, this additional contact interaction may be ignored in the Schrödinger equation if we adopt the conventional Aharonov-Bohm boundary condition that the center-of-mass wavefunction vanishes at the origin:

$$\Psi(\vec{x})\,|_{\text{origin}} = 0 \qquad (6.49)$$

Thus, the center-of-mass 2-particle Schrödinger equation is

$$\left(\frac{1}{r}\frac{d}{dr}r\frac{d}{dr} - \frac{(l+\alpha)^2}{r^2} + k^2 \right) \Psi = 0 \qquad (6.50)$$

where l denotes the angular momentum in the plane, and where α is related to the Chern-Simons coupling κ by

$$\alpha = \frac{1}{2\pi\kappa} \tag{6.51}$$

The scattering amplitude $f(k,\theta)$ is defined, as usual, by the asymptotic $(r \to \infty)$ form of the solution

$$\Psi(\vec{x}) \sim e^{ikr\cos\theta} + \frac{f(k,\theta)}{\sqrt{r}} e^{i(kr+\pi/4)} \tag{6.52}$$

This scattering amplitude was computed exactly by Aharonov and Bohm [2] - for the scattering of two identical particles (and with $|\alpha| < 1$) we have

$$f_{\text{total}}(k,\theta) = -\frac{i}{\sqrt{\pi k}} \sin(\pi\alpha) \left[\cot\theta - i\frac{\alpha}{|\alpha|} \right] \tag{6.53}$$

where $\theta \neq 0, \pi$. (For $\theta = 0, \pi$ there is an additional contribution which is needed to verify the unitarity of the scattering matrix [215,13]). Note that the scattering amplitude (6.53) only depends on the momentum k in the overall *kinematic* factor, which is a clear signal of scale invariance. Since this *exact* quantum mechanical answer is scale invariant, this provides strong motivation for developing a perturbative expansion which is also scale invariant. *A posteriori*, this is the reason for incorporating the contact interaction into the field theoretic perturbative analysis - such a term ensures that scale invariance is maintained.

For small α, the exact scattering amplitude (6.53) may be expanded as

$$f_{\text{total}}(k,\theta) = -i\alpha\sqrt{\frac{\pi}{k}} \left[\cot\theta - i\frac{\alpha}{|\alpha|} \right] + 0(\alpha^3) \tag{6.54}$$

while the naive Born approximation applied to the Schrödinger equation (6.50) yields

$$f_{\text{total}}(k,\theta) = -i\alpha\sqrt{\frac{\pi}{k}} \cot\theta + 0(\alpha^2) \qquad , \theta \neq 0, \pi \tag{6.55}$$

The nonanalytic $\frac{\alpha}{|\alpha|}$ term is absent in the leading Born approximation, and the $0(\alpha^2)$ term is in fact divergent. This divergence may be traced to the singular nature of the s-wave potential

$$\frac{\alpha^2}{r^2} \tag{6.56}$$

in the 2-body Schrödinger equation (6.50). This, in turn, is associated with the fact that the unperturbed s-wave wavefunctions do not satisfy the Aharonov-Bohm boundary condition (6.49) of vanishing at the origin. We consider this is more detail below.

In a field theoretical treatment, these divergences may be handled by regularization and renormalization. Bergman and Lozano showed that in order to develop a consistent perturbative renormalization it is necessary to introduce a quartic "contact interaction" term

$$+\frac{g}{2}|\psi|^4 \tag{6.57}$$

into the bare Lagrange density. The perturbatively generated infinities may then be absorbed into redefinitions of the bare coupling g. This procedure may be implemented explicitly either by a momentum cut-off or by using dimensional regularization together with a minimal subtraction. Each approach leads to a *renormalized* total scattering amplitude [21,13]

$$f_{\text{total}}^{(ren)}(k,\theta;\mu) = \frac{1}{\sqrt{4\pi k}}\left(-\frac{i}{\kappa}cot\theta + g + \frac{1}{\pi}\left(g^2 - \frac{1}{\kappa^2}\right)\left(ln\frac{\mu^2}{k^2} + i\pi\right)\right) \tag{6.58}$$

where μ is an arbitrary but necessary renormalization scale, and where g denotes the *renormalized* quartic coupling. Note that the presence of μ indicates that the renormalized scattering amplitude is *not* scale invariant, reflecting the quantum breaking of the classical scale invariance. However, at the special *self-dual* coupling point,

173

$$g = \pm \frac{1}{|\kappa|} \tag{6.59}$$

this scale dependence disappears, producing a scale invaraint result. Furthermore, the renormalized scattering amplitude is then

$$f_{\text{total}}^{(ren)}(k, \theta) = -i\alpha \sqrt{\frac{\pi}{k}} \left[cot\theta \pm i\frac{\alpha}{|\alpha|} \right] \tag{6.60}$$

This agrees exactly with the perturbative expansion (6.54) of the exact Aharonov-Bohm result, provided the contact interaction is chosen with the lower sign choice in (6.59). This choice corresponds to a *repulsive* delta-function interaction. This may be thought of crudely as an explicit implementation of the "hard-core" boundary condition (6.49) within the perturbative expansion.

To summarize, the field theoretic analysis of Bergman and Lozano shows that the difficulties with the perturbative expansion of Aharonov-Bohm scattering may be resolved by the introduction of a contact interaction. This makes the theory perturbatively renormalizable. Furthermore, if this contact interaction has the self-dual magnitude (6.59), then scale invariance is restored at the quantum level (to one-loop); and if the *sign* of the contact interaction is taken to correspond to repulsive delta-funcion interactions, then the renormalized scattering amplitude agrees with the perturbative expansion of the exact result. This result was extended through three-loop order using the technique of differential regularization and renormalization [87]. More recently, Kim has shown that this result indeed extends to *all orders* in perturbation theory [153]. In effect, the divergences that arise in the perturbative expansion for Aharonov-Bohm scattering are cancelled order by order by divergences coming from the contact interaction. These cancellations only occur at the self-dual point, and correspond to an all orders restoration of the classical scale invariance. At this self-dual point the theory is in fact

174

finite. Bak and Bergman have extended the one-loop analysis to the nonabelian case, using nonabelian nonrelativistic Chern-Simons field theory [13]. Here again, at the self-dual point the theory is finite and scale invariant, with the *repulsive* contact interaction reproducing the Aharonov-Bohm scattering amplitude. Interestingly, they find that the quartic contact interaction considered in Chapter 3 may be generalized slightly to allow for different $\pm$ signs in the various irreducible algebraic blocks.

We also note here that the *fermionic* Aharonov-Bohm scattering problem may be treated similarly in a field theoretic formulation, but in that case the contact interaction is unnecessary [21]. This follows because the fermionic theory includes a Pauli interaction term

$$\pm \frac{1}{2m} B \psi^\dagger \psi \qquad (6.61)$$

which corresponds exactly to the self-dual contact interaction term since $B = \frac{1}{\kappa} \psi^\dagger \psi$. The $\pm$ sign in (6.61) may be thought of as the sign of the Pauli spin projection or equivalently as the choice of self-duality or anti-self-duality.

To conclude this discussion of Aharonov-Bohm scattering, we comment on the first-quantized interpretation of these results. While the above results were derived in a second-quantized field theoretic formulation, there is of course a corresponding first-quantized quantum mechanical analysis. The key to this correspondence is the relationship between the contact interaction and the coincident-point boundary conditions imposed on the multi-particle wavefunctions. In the study of the anyonic many-body problem it has long been appreciated that the singular α^2/r^2 potential leads to divergences. To illustrate this, it is instructive to consider the problem of two anyons in an external harmonic oscillator well. This is an exactly solvable model, just as the Aharonov-Bohm scattering is a solvable two-particle problem. Furthermore, the introduction of the external harmonic well provides a physical regularization of the free system that is extremely useful in the computation of various statistical mechanical properties of the anyon gas [191,43,45,37,194]. It is also important because the harmonic well may readily be adapted to the phenomenologically interesting model of anyons in an external magnetic field [175]. The center-of-mass first-quantized Hamiltonian operator is

$$H = \frac{1}{2}\left(-\frac{1}{r}\frac{\partial}{\partial r}\left(r\frac{\partial}{\partial r}\right) - \frac{1}{r^2}\frac{\partial}{\partial \theta^2} + r^2 - \frac{2i\alpha}{r^2}\frac{\partial}{\partial \theta} + \frac{\alpha^2}{r^2}\right) \qquad (6.62)$$

The exact eigenstates of this Hamiltonian are

$$\Psi_{n,l} = r^{|l+\alpha|}e^{il\theta}e^{-r^2/2}L_n^{|l+\alpha|}(r^2) \qquad (6.63)$$

where the L_n^j are Laguerre polynomials. The corresponding energy eigenvalues are

$$E_{n,l} = 2n + |l + \alpha| + 1 \qquad (6.64)$$

with $n = 0, 1, 2, \ldots$, and l is an integer such that $|l + \alpha| \geq -n$. The principal difficulty with a perturbative analysis of this problem arises in the s-wave ($l = 0$) equation, for which the Schrödinger equation becomes an ordinary differential equation. The presence of the

$$\frac{\alpha^2}{r^2} \qquad (6.65)$$

potential term makes the origin $r = 0$ an irregular singular point. (Note that since we are in the center-of-mass frame, the origin corresponds to the point where the two particles coincide.) This term may be removed from the ODE by a redefinition of the wavefunction

$$\Psi = r^{\pm|\alpha|}\tilde{\Psi} \qquad (6.66)$$

The transformed (s-wave) hamiltonian operator for $\tilde{\Psi}$ now reads

$$\tilde{H} = \frac{1}{2}\left(-\frac{1}{r}\frac{\partial}{\partial r}\left(r\frac{\partial}{\partial r}\right) + r^2 \pm |\alpha|\frac{1}{r}\frac{\partial}{\partial r}\right) \qquad (6.67)$$

One may then develop a perturbative expansion for the Schrödinger equation satisfied by $\tilde{\Psi}$ and $\tilde{H}$, including the non-hermitean term $\pm|\alpha|\frac{1}{r}\frac{\partial}{\partial r}$. This approach has been tested thoroughly and has led to extensive systematic results for the perturbative analysis of the multi-anyon problem [43,45,37]. Another useful perturbative approach [34,35] involves explicitly making the unperturbed wavefunctions hard-core. For further discussion of the relations between these various approaches see [6].

An alternative (and presumably equivalent - although the correspondence has only been investigated explicitly for various simple but illustrative examples) regularization procedure for the multi-anyon problem is the introduction of delta-function potential interactions corresponding precisely to the field theoretic quartic contact interaction [6,7]. (This has also been successfully tested in a perturbative *field theoretical* study of the anyon gas through three-loop order [236,77].) Thus, we consider

$$H^{(\alpha)} = H \mp 2\pi|\alpha| \sum_{I<J} \delta(\vec{x}_I - \vec{x}_J) \qquad (6.68)$$

For "hard-core" anyons, with wavefunctions vanishing at coincident points, this added interaction should (formally) have no effect on the *exact* wavefunctions. Nevertheless, these additional interaction terms have a significant (positive) effect on the perturbative expansion. For example, the two-particle Hamiltonian becomes

$$H \to H \mp 2\pi|\alpha|\delta(\vec{x}) \qquad (6.69)$$

and the delta-function potential leads to divergences which exactly cancel the divergences from the α^2/r^2 term. This leads to finite results for the "bosonic end" perturbation about $\alpha = 0$. This is precisely the quantum mechanical realization of the Bergman-Lozano contact interaction regularization in the field theoretic analysis of Aharonov-Bohm scattering. The choice of the coefficient $\mp 2\pi|\alpha|$ in the delta function terms in (6.69) corresponds to the self-dual coupling $g = \pm 2\pi|\alpha| = \pm 1/|\kappa|$ of the quartic contact interaction. Furthermore, as in the Aharonov-Bohm case, the *sign* of this term can only be decided by comparing with an exact result. Comparing with the exact energy eigenvalues (6.64) for the two-body system we see that we require the lower sign choice, which corresponds to a *repulsive* delta-function interaction. By considering the two-body Schrödinger equation more carefully, we see that this sign in E_{nl} is determined by the boundary conditions imposed on the exact wavefunctions $\Psi_{n,l}$ - *viz.* that the exact wavefunctions vanish as $r \to 0$. Thus, the solutions going like $r^{-|l+\alpha|}$ at the origin were rejected.

This shows a direct and precise link between the choice of boundary conditions in the first-quantized problem and the choice of sign for the delta-function interaction, which translates into the choice of sign for the $|\psi|^4$ contact potential term in the nonrelativistic Chern-Simons field theory. For the Aharonov-Bohm scattering system, and for the system of two anyons in a harmonic well, the standard boundary conditions translate into a *positive* coefficient for the $|\psi|^4$ term in the self-dual Chern-Simons Hamiltonian. It is interesting to note that this is *opposite* to the sign choice made previously (see Chapters 2 and 3) in our study of the classical nonrelativistic self-dual Chern-Simons systems. There the choice of sign was also based on a choice of boundary conditions. The sign correlation was chosen in order to obtain regular, integrable and finite solutions of the Liouville equation (2.77). As discussed in Section II E, it is also possible to find solutions to the Liouville equation with the opposite sign correlation, but then these solutions are necessarily singular at some point(s).

This strange inversion of boundary conditions may also be seen [7] in terms of a self-adjoint extension [4,215,97] of the quantum mechanical Hamiltonian. For example, for the 2-particle Aharonov-Bohm scattering problem we may consider instead the scattering problem on the plane with a point (corresponding to the coincident point) removed, but with a one-parameter family of boundary conditions specified at this point for the s-wave wavefunctions. In this spirit, Manuel and Tarrach [187–189] have shown that the suitable one-parameter family of boundary conditions may be expressed as

$$\Psi(\vec{x}) \sim r^{|\alpha|} + w \left(\frac{r_0^2}{r} \right)^{|\alpha|} \tag{6.70}$$

for $r \to 0$. Here, r_0 is some arbitrary reference length scale and w is the dimensionless (real) self-adjoint extension parameter. The appearance of the reference length scale r_0 breaks the classical scale invariance, but this scale invariance may be restored in the limit

$$w = 0 \qquad \text{or} \qquad w = \infty \tag{6.71}$$

The first choice, $w = 0$, corresponds to the standard hard-core boundary condition, while the latter $w = \infty$, corresponds to wavefunctions which diverge at coincident points. Amelino-Camelia and Bak [7] have translated these boundary conditions into the corresponding contact interaction, and found that the strength of the $|\psi|^4$ contact interaction is

$$g = 2\pi |\alpha| \frac{1 - w}{1 + w} \tag{6.72}$$

Thus, taking w to 0 and to ∞, we see that the two critical couplings which restore scale invariance are $g = \pm 2\pi |\alpha|$, as before. This unusual inversion of boundary conditions is a significant unexplained fact which promises to elucidate the role of the self-dual Chern-Simons solitons in the quantum theory.

D. Quantum Aspects of Relativistic SDCS Theories

The current situation with quantum relativistic self-dual Chern-Simons models is not as well developed as for the corresponding non-relativistic systems. Numerous interesting results have been found, and yet there remain many more challenging and important problems. In this concluding Section I give a very brief summary of the current state of affairs, which I hope may be useful as a guide to further developments.

One of the most obvious, and most pressing, questions is to ask what happens to the self-dual structure of the theory once it is quantized. Is the self-duality preserved? If not, what breaks the self-duality? And, if self-duality *is* preserved, one must ask *why* it is preserved. In the nonrelativistic models (as discussed in the previous few Sections) the self-duality condition in the quantum theory is tied to the preservation of scale invariance in the quantum theory. In the relativistic systems there is no such scale symmetry, even at the classical level. Rather, the self-duality condition is tied to the existence of an $N = 2$ supersymmetry (see Section IV F). It is *this* symmetry that is preserved at the quantum level as a consequence of the self-dual form of the Lagrange density. Evidence for this can be found, for example, in a comprehensive renormalization group analysis of Chern-Simons-matter models [9,10] (see also studies of the effective potential [119,162]). These models are formally renormalizable by standard power counting arguments, and a minimal subtraction dimensional regularization scheme has been used to perform two-loop computations (to this order there are no ambiguities in using dimensional regularization to compute beta functions and renormalization group flows). This work has shown that while the gauge coupling is not infinitely renormalized, the matter couplings have intricate renormalization group flows. The self-dual couplings corresponding to $N = 2$ supersymmetry appear as a fixed point, and this fixed point has interesting stability properties which depend on the gauge group. Given that there is a natural nonrelativistic limit linking the relativistic and nonrelativistic self-dual Chern-Simons models at the classical level, it would be interesting to understand this complicated phase structure of the relativistic model in terms of its nonrelativistic reduction to a theory which supports scale symmetry at the quantum level. Indeed, one might even expect that the quantum supersymmetry algebra would reduce in a nonrelativistic limit to a symmetry algebra generating scale transformations. This has not been demonstrated explicitly, although the nonrelativistic super-Galilean structure of the nonrelativistic self-dual Chern-Simons systems has been discussed in [161].

This renormalization group analysis is related to the question of the renormalization of the couplings in the self-dual Chern-Simons system. This is of particular interest because Chern-Simons theories are well known to have special renormalization properties. For example, for abelian Chern-Simons theories the Coleman-Hill theorem [36] states that in a large class of $2 + 1$ dimensional matter-gauge theories, the Chern-Simons coupling parameter is not renormalized, except by fermionic contributions at one loop. This theorem does not cover spontaneously broken theories, but this result, when rephrased in terms of the effective action, has since been extended to incorporate spontaneously broken theories [148,144]. This is important, as the self-dual potential of the self-dual Chern-Simons model possesses distinct vacua and acts as a symmetry breaking potential. Nevertheless, the renormalization of the abelian $N = 2$ and $N = 3$ supersymmetric Chern-Simons Higgs models has been discussed in [143], where it is shown that the Chern-Simons coupling coefficient κ receives a finite shift which is a multiple of the basic fermionic contribution $1/4\pi$.

Such a shift would be even more significant in the nonabelian theory, because in the nonabelian theory κ itself must be quantized in integer multiples of $1/4\pi$ to preserve invariance under topologically nontrivial gauge transformations [46]. In the absence of matter fields, in a nonabelian theory κ receives a finite shift [205,249] which is itself a multiple of $1/4\pi$, thereby preserving the quantum consistency condition. This has also been shown [33,147] to be true in the broken phase of a spontaneously broken Chern-Simons-Yang-Mills-Higgs theory when the broken phase possesses a residual nonabelian symmetry (whose consistency must be protected). This has recently been studied [145] in the *supersymmetric* Yang-Mills Chern-Simons theories [141,142], with the interesting result that in the $N = 2$ and $N = 3$ SUSY theories the bosonic and fermionic sector shifts cancel one another exactly. Presumably, a similar mechanism operates in the self-dual Chern-Simons theories in which the Yang-Mills term is absent, although the computations are somewhat different since the Chern-Simons-Higgs mechanism acts in a different manner in this case (see Section IV E). Furthermore, the self-dual potential has a very special form and its overall strength is determined by the Chern-Simons coupling coefficient. Thus, for self-duality to be preserved after renormalization, the matter fields and couplings must be renormalized in a precisely corresponding manner. These issues, together with the related issues of vacuum tunnelling and mass resonances, deserve to be explored in more detail.

Another significant line of investigation concerns the application to self-dual Chern-Simons theories of Manton's technique [186] for analyzing the slow motion of vortices. In this approach one constructs an effective quantum mechanical Lagrangian which describes the fluctuations about the static self-dual classical configurations.

As the static configurations saturate the classical Bogomol'nyi lower bound on the energy, one assumes that the slow motion of vortices can only lead to 'small' deformations of the classical fields with respect to the self-dual configurations. This method was originally developed for the BPS monopole system [186], and later applied to the Abelian Higgs model [214,216], so it is natural to consider its extension to the self-dual Chern-Simons systems. For the *topological* vortices, there is an integer topological index N which may be interpreted as the number of vortices associated with the given self-dual solution. The general vortex solution may be characterized by $2N$ parameters, $\vec{z}_i$ $(i = 1 \ldots N)$, which may be chosen to correspond to the zeros of the scalar field ϕ [127,129,240]. If these zeros are widely separated, then they may be identified with the locations of the spatially separated vortices. Allowing the parameters $\vec{z}_i$ to become time dependent, and integrating over the spatial coordinates with the general static self-dual solutions, leads to an effective Lagrangian for the fluctuations of these vortices about their self-dual configuration. For the 'slow' motion of vortices (*i.e.* when $|\dot{\vec{z}}_i|$ is small in some appropriate sense) it is possible to identify kinetic and potential contributions to this effective quantum mechanical Lagrangian. Thus, the infinite dimensional field theory is (formally) reduced to a finite dimensional parameter space, whose dynamics is determined by the self-dual Chern-Simons solitons. Simliar techniques have been applied to kinematical issues in the quantization of the nonrelativistic self-dual Chern-Simons systems [139,116,183].

For the relativistic abelian self-dual Chern-Simons system, Kim and Min [152] showed that in the 'slow motion limit' this effective Lagrangian contains the expected kinetic term for N point particle vortices, with the correct masses, as well as an anyonic statistical interaction which reflects the known anyonic character of the self-dual Chern-Simons vortices [127]. In other cases [186,216,217] Manton's technique leads to a beautiful geometrical picture in terms of geodesics on the finite dimensional manifold of self-dual configurations with a given topological index. This, in turn, has important consequences for the scattering of such vortices [186,214]. However, for the self-dual Chern-Simons systems this geometrical construction and the associated scattering theory have not yet been resolved. The major complication is the presence in the Lagrangian of terms that are first-order in time derivatives. Physically, this reflects the appearance of velocity dependent forces. Such velocity dependent forces are, in fact, crucial for the consistency of the relativistic self-dual Chern-Simons models. Y. Kim and K. Lee have shown [155] that the velocity dependent Magnus force (which operates like a Lorentz force in the dual formulation) is necessary in order to resolve an apparent inconsistency in the spin-statistics relation for the self-dual Chern-Simons vortices. The correct spin-statistics relation requires a Magnus force between vortices which leads to an additional Aharonov-Bohm phase, in addition to that coming from the electromagnetic interactions of the vortices. These effects are elegantly described in a dual formalism in which the asymmetric phase massive gauge excitation is characterized by a Maxwell-Chern-Simons Lagrangian rather than a Chern-Simons-Higgs Lagrangian. In this dual formulation the vortices appear as charged particles and both the original electromagnetic force and the velocity dependent Magnus force become naturally unified into a single dual gauge force from which the correct spin-statistics relation follows [155]. For some implications of these issues for scattering of Chern-Simons vortices see [73,74].

A closely related quantum investigation concerns the quantization of the zero modes about the classical vortex solutions. The bosonic zero modes have been counted [129,167] and applied to the collective modes of the solitons [155]. In the supersymmetric self-dual Chern-Simons theories there are also fermionic fields and these possess zero modes about the classical vortices. (These fermionic zero modes are also related to the scattering of fermions from a self-dual Chern-Simons vortex [101].) Owing to the supersymmetry, these fermionic zero modes are of course closely related to the bosonic ones. For the $N = 2$ supersymmetric model (which corresponds to the self-dual model described in Chapter 4) the number of bosonic and fermionic zero modes about a self-dual Chern-Simons vortex are equal. Furthermore [164], the bosonic and fermionic density of modes, and the spectra themselves, are equal. This has the interesting consequence that when these mode fluctuations are quantized to compute the leading mass correction to the vortex mass, the bosonic and fermionic contributions cancel exactly. Thus, in the abelian $N = 2$ supersymmetric model there is no leading order quantum mass correction to the self-dual Chern-Simons vortex mass [164]. This analysis also reveals the supermultiplet structure of the self-dual Chern-Simons vortices, although it is not yet clear whether the Bogomol'nyi bound, relating the vortex mass to the magnetic flux, is actually saturated at the quantum level. This is a general feature of other self-dual systems [248,111,112] and it would be interesting to find a detailed understanding of how this works in the self-dual Chern-Simons models.

The possibility of a complete, explicit quantization of the relativistic self-dual Chern-Simons theories is inherently linked to the question of the integrability and/or solvability of the associated self-duality equations. Since both the Chern-Simons and Abelian Higgs models have a Bogomol'nyi structure for their respective self-dual potentials, it is natural to consider similarities (and differences) between these two models. In each case one may deduce, and prove, the existence of vortex solutions; although no exact solutions are known in either case. Using complex analytical and variational techniques (similar to methods used in the study of Nielsen-Olesen vortices [137]), Wang has shown [240] that the abelian self-dual Chern-Simons system has topological multi-vortex solutions which may be characterized by a finite set of zeros of the scalar field ϕ. The net vortex number is the sum of the orders of these zeros, and the zeros may be identified with the locations of the vortices. Spruck and Yang have presented an efficient constructive and numerical method for generating such multi-vortex solutions, and studying their properties [226]. They have also proven the existence of radially symmetric nontopological vortex solutions, corresponding to vortices superimposed at a point [225]. In the absence of exact closed-form solutions, the question of integrability must be phrased in precise physical terms, as 'integrability' in the absence of any obvious time evolution (with associated conserved quantities etc...) is rather difficult to define [241–243,255]. Schiff has analyzed the self-duality equations for the self-dual Chern-Simon system and for the abelian Higgs model, using the Painlevé criterion for integrability [218]. This condition defines integrability in terms of the existence and classification of movable singularities of the differential equation. This Painlevé analysis was performed for both the ordinary differential equations which result from restricting to radial solutions, and for the full two-dimensional partial differential equations. Both the self-dual Chern-Simons and the abelian Higgs model are nonintegrable (in the Painlevé sense) for radial solutions with a flat spatial metric, but there exists a special choice of metric which yields explicit radial solutions in the Chern-Simons case. However, the physical significance of these special solutions is unclear. For the partial differential

186

equations, the Chern-Simons system is nonintegrable for any spatial metric, while for the abelian Higgs there exist some metrics which lead to integrable equations [218]. A Painlevé analysis of the full $2+1$ dimensional equations of motion of the *nonrelativistic* abelian self-dual Chern-Simons model (discussed in Chapter II) has recently been performed [156] (see also [176]), with strong indications that the full time-dependent system of equations is *not* completely integrable. An interesting line of investigation concerns the reduction of certain nonlinear σ-models to the self-dual Chern-Simons systems [190,199]. The full implications of these various integrability studies for the quantum theory have yet to be explored in detail, and deserve to be pursued further.

Many of the issues described in this last Section are still under active current investigation, and I hope these Lecture Notes may be of some help as a guide for future developments.

VII. BIBLIOGRAPHY

[1] I. Affleck, J. Harvey and E. Witten, "Instantons and (Super-) Symmetry Breaking in (2+1) Dimensions", *Nucl. Phys. B* **206** (1982) 413.

[2] Y. Aharonov and D. Bohm, "Significance of Electromagnetic Potentials in the Quantum Theory", *Phys. Rev.* **115** (1959) 485.

[3] Y. Aharonov, C. K. Au, E. Lerner and J. Q. Liang, "Aharonov-Bohm Effect as a Scattering Event", *Phys. Rev. D* **29** (1984) 2396.

[4] S. Albeverio et al, *Solvable Models in Quantum Mechanics*, (Springer-Verlag, Berlin, 1988).

[5] G. Amelino-Camelia, C. Chou and L. Hua, "Perturbative Anyon Spectra from the Bosonic End", *Phys. Lett. B* **286** (1992) 329.

[6] G. Amelino-Camelia, "Perturbative Renormalizations of Anyon Quantum Mechanics", *Phys. Rev. D* **51** (1995) 2000.

[7] G. Amelino-Camelia and D. Bak, "Schrödinger Self-Adjoint Extension and Quantum Field Theory", *Phys. Lett. B* **343** (1995) 231.

[8] A. Antillón, J. Escalona, G. Germán and M. Torres, "Self-Dual Nonabelian Vortices in a Φ^2 Chern-Simons Theory", UNAM-Mexico preprint, May 1995; hep-th/9505115.

[9] L. Avdeev, G. Grigoryev and D. Kazakov, "Renormalizations in Abelian Chern-Simons Field Theories with Matter", *Nucl. Phys. B* **382** (1992) 561.

[10] L. Avdeev, D. Kazakov and I. Kondrashuk, "Renormalizations in Supersymmetric and Nonsupersymmetric Nonabelian Chern-Simons Field Theories with Matter", *Nucl. Phys. B* **391** (1993) 333.

[11] S. Axelrod and I. Singer, "Chern-Simons Perturbation Theory", in Proceedings of XXth *International Conference on Differential Geometric Methods in Theoretical Physics*, (New York 1991), S. Catto and A. Rocha, Eds. (World Scientific, 1992).

[12] D. Bak, R. Jackiw and S-Y. Pi, "Non-Abelian Chern-Simons Particles and Their Quantization", *Phys. Rev. D* **49** (1994) 6778.

[13] D. Bak and O. Bergman, "Perturbative Analysis of Nonabelian Aharonov-Bohm Scattering", *Phys. Rev. D* **51** (1995) 1994.

[14] I. Barashenkov and A. Harin, "Nonrelativistic Chern-Simons Theory for the Repulsive Bose Gas", *Phys. Rev. Lett.* **72** (1994) 1575.

[15] C. Baxter, "Cold Rydberg Atoms as Realizable Analogs of Chern-Simons Theory", *Phys. Rev. Lett.* **74** (1995) 514.

[16] D. Bazeia and G. Lozano, "Nontopological Solitons in Chern-Simons Systems", *Phys. Rev. D* **44** (1991) 3348.

[17] M. A. B. Beg and R. Furlong, "$\lambda\phi^4$ Theory in the Nonrelativistic Limit", *Phys. Rev. D* **31** (1985) 1370.

[18] C. Bender and G. Dunne, "Covariance of the Gauge Field in Three Dimensional Quantum Electrodynamics", *Phys. Rev. D* **44** (1991) 2565.

[19] L. Bergé, A. de Bouard and J. Saut, "Collapse of Chern-Simons-Gauged Matter Fields", *Phys. Rev. Lett.* **74** (1995) 3907.

[20] O. Bergman, "Non Relativistic Field Theoretic Scale Anomaly", *Phys. Rev. D* **46** (1992) 5474.

[21] O. Bergman and G. Lozano, "Aharonov-Bohm Scattering, Contact Interactions and Scale Invariance", *Ann. Phys.* **229** (1994) 416.

[22] B. Binegar, "Relativistic Field Theories in Three Dimensions", *J. Math. Phys.* **23** (1982) 1511.

[23] D. Birmingham, M. Blau, M. Rakowski and G. Thompson, "Topological Field Theory", *Phys. Rep.* **209** (1991) 129.

[24] E. Bogomol'nyi, "Stability of Classical Solutions", *Sov. J. Nucl. Phys* **24** (1976) 449.

[25] M. Bos and V. P. Nair, "Coherent State Quantization of Chern-Simon Theory", *Int. J. Mod. Phys. A* **5** (1990) 959.

[26] N. Bourbaki, *Groupes et Algèbres de Lie*, *VI* (Hermann, Paris, 1968).

[27] D. Caenepeel, F. Gingras, M. Leblanc and D. McKeon, "Structure of the Effective Potential in Nonrelativistic Chern-Simons Field Theory", *Phys. Rev. D* **49** (1994) 5422.

[28] D. Caenepeel and M. Leblanc, "Effective Potential for Nonrelativistic Nonabelian Chern-Simons Matter System in Constant Background Fields", Montréal preprint CRM-2195 (June 1994); hep-th/9406187.

[29] D. Cangemi and C. Lee, "Self-Dual Chern-Simons Solitons in (2+1)-Dimensional Einstein Gravity", *Phys. Rev. D* **46** (1992) 4768.

[30] D. Cangemi, "Self-Dual Chern-Simons Solitons with Noncompact Groups", *J. Phys. A: Math. and Gen.* **26** (1993) 2945.

[31] A. Cappelli, C. A. Trugenberger and G. Zemba, "Stable Hierarchical Quantum Hall Fluids as $W_{1+\infty}$ Minimal Models", *Nucl. Phys. B* **448** (1995) 470.

[32] R. Carter, *Simple Groups of Lie Type* (Wiley, New York 1972).

[33] L. Chen, G. Dunne, K. Haller and E. Lim-Lombridas, "Integer Quantization of the Chern-Simons Coefficient in a Broken Phase", *Phys. Lett. B* **348** (1995) 468.

[34] C. Chou, "The Multi-Anyon Spectra and Wavefunctions", *Phys. Rev. D* **44** (1991) 2533, (E) **45** (1992) 1433.

[35] C. Chou, L. Hua and G. Amelino-Camelia, "Perturbative Anyon Spectra from the Bosonic End", *Phys. Lett. B* **286** (1992) 329.

[36] S. Coleman and B. Hill, "No More Corrections to the Topological Mass Term in QED_3", *Phys. Lett. B* **159** (1985) 184.

[37] A. Comtet, J. McCabe and S. Ouvry, "Perturbative Equation of State for a Gas of Anyons", *Phys. Lett. B* **260** (1991) 372.

[38] F. Cooper, A. Khare and U. Sukhatme, "Supersymmetry and Quantum Mechanics", *Phys. Rep.* **251** (1995) 267.

[39] E. Corrigan, D. Fairlie, J. Nuyts and D. Olive, "Magnetic Monopoles in $SU(3)$ Gauge Theories", *Nucl. Phys. B* **106** (1976) 475.

[40] E. Corrigan, "Recent Developments in Affine Toda Quantum Field Theory", Lectures at CRM-CAP Summer School on Particles and Fields '94, Banff, Canada, 16-24 Aug 1994; hep-th/9412213.

[41] L. Cugliandolo, G. Lozano, M. Manías and F. Schaposnik, "Bogomol'nyi Equations for Nonabelian Chern Simons Higgs Theories", *Mod. Phys. Lett. A* **6** (1991) 479.

[42] A. Das, *Integrable Models* (World Scientific 1989).

[43] A. Dasnières de Veigy and S. Ouvry, "Perturbative Anyon Gas", *Nucl. Phys.* **B388** (1992) 718.

[44] A. Dasnières de Veigy and S. Ouvry, "Topological Two Dimensional Quantum Mechanics", *Phys. Lett. B* **307** (1993) 91.

[45] A. Dasnières de Veigy and S. Ouvry, "Equation of State of an Anyon Gas in a Strong Magnetic Field", *Phys. Rev. Lett.* **72** (1994) 600.

[46] S. Deser, R. Jackiw and S. Templeton, "Topologically Massive Gauge Theory", *Ann. Phys. (NY)* **140** (1982) 372.

[47] S. Deser and R. Jackiw, "Self-Duality of Topologically Massive Gauge Theories", *Phys. Lett. B* **139** (1984) 371.

[48] S. Deser and Z. Yang, "A Comment on the Higgs Effect in Presence of Chern-Simons Terms", *Mod. Phys. Lett. A* **3** (1989) 2123.

[49] H. J. de Vega and F. Schaposnik, "Electrically Charged Vortices in Nonabelian Gauge Theories with Chern-Simons Term", *Phys. Rev. Lett.* **56** (1986) 2564.

[50] P. Dirac, "Quantized Singularities in the Electromagnetic Field", *Proc. Roy. Soc.* **A133** (1931) 60.

[51] P. Dirac, "The Theory of Magnetic Poles", *Phys. Rev.* **74** (1948) 817.

[52] S. Donaldson, "Twisted Harmonic Maps and the Self-Duality Equations", *Proc. Lond. Math. Soc.* **55** (1987) 127.

[53] P. Donatis and R. Iengo, "Comment on Vortices in Chern-Simons and Maxwell Electrodynamics", *Phys. Lett. B* **320** (1994) 64.

[54] G. Dunne, R. Jackiw and C. Trugenberger, "Chern-Simons Theory in the Schrödinger Representation", *Ann. Phys.* **194** (1989) 197.

[55] G. Dunne and C. Trugenberger, "Odd Dimensional Gauge Theory and Current Algebra", *Ann. Phys.* **204** (1990) 281.

[56] G. Dunne, R. Jackiw and C. Trugenberger, "Topological (Chern-Simons) Quantum Mechanics", *Phys. Rev. D* **41** (1990) 661.

[57] G. Dunne and C. Trugenberger, "Self-Duality and Non-Relativistic Maxwell-Chern-Simons Solitons", *Phys. Rev. D* **43** (1991) 1323.

[58] G. Dunne, R. Jackiw, S-Y. Pi and C. Trugenberger, " Self-Dual Chern-Simons Solitons and Two-Dimensional Nonlinear Equations", *Phys. Rev. D* **43** (1991) 1332, (E) **45** (1992) 3012.

[59] G. Dunne, "Chern-Simons Solitons, Toda Theories and the Chiral Model", *Comm. Math. Phys.* **150** (1992) 519.

[60] G. Dunne and R. Jackiw, " 'Peierls Substitution' and Chern-Simons Quantum Mechanics", *Nucl. Phys. B (Proc. Suppl.)* **33C** (1993) 114.

[61] G. Dunne, "Classification of Nonabelian Chern-Simons Vortices", in *Proceedings of XXIInd International Conference on Differen-*

194

tial Geometric Methods in Physics, Ixtapa (Mexico) 1993; hep-th/9310182, published in *Adv. in Applied Clifford Algebras (Proc. Suppl.)* **4** (1994) 229-238.

[62] G. Dunne, "Relativistic Self-Dual Chern-Simons Vortices with Adjoint Coupling", *Phys. Lett. B* **324** (1994) 359.

[63] G. Dunne, "Symmetry Breaking in the Schrödinger Representation for Chern-Simons Theories", *Phys. Rev. D* **50** (1994) 5321.

[64] G. Dunne, "Vacuum Mass Spectra for $SU(N)$ Self-Dual Chern-Simons-Higgs Systems", *Nucl. Phys. B* **433** (1995) 333.

[65] G. Dunne, "Mass Degeneracies in Self-Dual Models", *Phys. Lett. B* **345** (1995) 452.

[66] G. Dunne, "Self-Dual Chern-Simons Theories", Lectures at 13th Symposium on Theoretical Physics, *Field Theory and Mathematical Physics*, Mt. Sorak, Korea, June-July 1994, published in the Proceedings, J. E. Kim, Ed. (Mineumsa, Seoul, 1995).

[67] C. Duval, P. Horvathy and L. Palla, "Conformal Symmetry of the Coupled Chern-Simons and Gauged Nonlinear Schrödinger Equations", *Phys. Lett. B* **325** (1994) 39.

[68] C. Duval, P. Horvathy and L. Palla, "Conformal Properties of Chern-Simons Vortices in External Fields", *Phys. Rev. D* **50** (1994) 6658.

[69] C. Duval, P. Horvathy and L. Palla, "Spinor Vortices in Non-relativistic Chern-Simons Theory", preprint March 1995; hep-th/9503061.

[70] E. Dynkin, "Semisimple Subalgebras of Semisimple Lie Algebras", *Amer. Math. Soc. Transl.* **6** (1957) 111.

[71] J. Dziarmaga, "Short-Range Interactions of Chern-Simons Vortices", *Phys. Lett. B* **320** (1994) 69.

[72] J. Dziarmaga, "Only Hybrid Anyons can Exist in Broken Symmetry Phase of Nonrelativistic $[U(1)]^2$ Chern-Simons Theory", *Phys. Rev. D* **50** (1994) R2376.

[73] J. Dziarmaga, "Low Energy Dynamics of $U(1)^N$ Chern-Simons Solitons", *Phys. Rev. D* **49** (1994) 5469.

[74] J. Dziarmaga, "More on Scattering of Chern-Simons Vortices", *Phys. Rev. D* **51** (1995) 7052.

[75] J. Edelstein, C. Núñez and F. Schaposnik, "Supersymmetry and Bogomol'nyi Equations in the Abelian Higgs Model", *Phys. Lett. B* **329** (1994) 39.

[76] S. Elitzur, G. Moore, A. Schwimmer and N. Seiberg, "Remarks on the Canonical Quantization of the Chern-Simons-Witten Theory", *Nucl. Phys. B* **326** (1989) 108.

196

[77] R. Emparan, M. Valle Basagoiti, "Three Loop Calculation of the Anyonic Full Cluster Expansion", *Mod. Phys. Lett.* **A8** (1993) 3291.

[78] Z. Ezawa, M. Hotta and A. Iwazaki, "Nonrelativistic Chern-Simons Vortex Solitons in External Magnetic Field", *Phys. Rev. D* **44** (1991) 452.

[79] Z. Ezawa, M. Hotta and A. Iwazaki, "Nonrelativistic Chern-Simons Vortices in Magnetic Field: Their Masses and Spins", *Phys. Lett.* **261B** (1991) 443.

[80] Z. Ezawa, M. Hotta and A. Iwazaki, "Breathing Vortex Solitons in Nonrelativistic Chern-Simons Gauge Theory", *Phys. Rev. Lett.* **67** (1991) 411.

[81] Z. Ezawa, M. Hotta and A. Iwazaki, "Time Dependent Topological Chern-Simons Solitons in External Magnetic Field", *Phys. Rev. D* **44** (1991) 3906.

[82] L. Faddeev and L. Takhtajan, *Hamiltonian Methods in the Theory of Solitons*, (Springer-Verlag, Berlin, 1987).

[83] L. Faddeev and R. Jackiw, "Hamiltonian Reduction of Constrained and Unconstrained Systems", *Phys. Rev. Lett.* **60** (1988) 1692.

[84] R. Floreanini, R. Percacci and E. Sezgin, "Infinite Dimensional Algebras in Chern-Simons Quantum Mechanics", *Phys. Lett. B* **261** (1991) 51.

[85] S. Forte, "Quantum Mechanics and Quantum Field Theory with Fractional Spin and Statistics", *Rev. Mod. Phys.* **64** (1992) 193.

[86] E. Fradkin, *Field Theories of Condensed Matter Systems*, (Addison-Wesley, Redwood City, 1991).

[87] D. Freedman, G. Lozano and N. Rius, "Differential Regularization of a Nonrelativistic Anyon Model", *Phys. Rev. D* **49** (1994) 1054.

[88] M. Freeman, "On the Mass Spectrum of Affine Toda Field Theory", *Phys. Lett. B* **261** (1991) 57.

[89] A. Fring, H. C. Liao and D. Olive, "The Mass Spectrum and Coupling in Affine Toda Theories", *Phys. Lett. B* **82** (1991) 82.

[90] J. Fröhlich and P. Marchetti, "Quantum Field Theory of Vortices and Anyons", *Comm. Math. Phys.* **121** (1989) 177.

[91] K. Fujii, "A Relation Between Instantons of Grassmann σ-Models and Toda Equations II", *Lett. Math. Phys.* **25** (1992) 203.

[92] K. Fujii, "Nonlinear Grassmann σ-Models, Toda Equations, and Self-Dual Einstein Equations: Supplements to Previous Papers", *Lett. Math. Phys.* **27** (1993) 117.

[93] N. Ganoulis, P. Goddard and D. Olive, "Self-Dual Monopoles and Toda Molecules", *Nucl. Phys. B* **205** **[FS]** (1982) 601.

[94] S. J. Gates, M. Grisaru, M. Rocek and W. Siegel, *Superspace* (Benjamin/Cummings 1983).

[95] S. J. Gates and H. Nishino, "Remarks on N=2 Supersymmetric Chern-Simons Theories", *Phys. Lett. B* **281** (1992) 72.

[96] L. Gendenshtein, "Supersymmetric Quantum Mechanics, the Electron in a Magnetic Field, and Vacuum Degeneracy", *Sov. J. Nucl. Phys.* **41** (1985) 166.

[97] P. de Sousa Gerbert, "Anyons, Chern-Simons Lagrangians and Physics in 2 + 1 Dimensions", *Int. Journ. Mod. Phys. A* **6** (1991) 173.

[98] S. Girvin and T. Jach, "Formalism for the Quantum Hall Effect: Hilbert Space of Analytic Functions", *Phys. Rev. B* **29** (1984) 5617.

[99] S. Girvin, A. MacDonald, M. Fisher, S-J. Rey and J. Sethna, "Exactly Soluble Model of Fractional Statistics", *Phys. Rev. Lett.* **65** (1990) 1671.

[100] P. Goddard and D. Olive, "Magnetic Monopoles in Gauge Field Theories", *Rep. Prog. Phys.* **41** (1978) 91.

[101] G. Grignani and G. Nardelli, "Scattering of Low-Energy Fermions by a Chern-Simons Vortex", *Phys. Rev. D* **42** (1990) 4145.

[102] G. Grigoryev and D. Kazakov, "Renormalization Group Study of Anyon Superconductivity", *Phys. Lett. B* **253** (1991) 411.

[103] B. Grossman, "Hierarchy of Soliton Solutions to the Gauged Nonlinear Schrödinger Equation on the Plane", *Phys. Rev. Lett.* **65** (1990) 3230.

[104] C. Hagen, "A New Gauge Theory Without an Elementary Photon", *Ann. Phys.* **157** (1984) 342.

[105] C. Hagen, "Galilean-Invariant Gauge Theory", *Phys. Rev. D* **31** (1985) 848.

[106] C. Hagen, "Comment on 'Soliton Solutions to the Gauged Nonlinear Schrödinger Equation on the Plane"', *Phys. Rev. Lett.* **66** (1991) 2681.

[107] C. Hagen, "Perturbation Theory and the Aharonov-Bohm Effect", Rochester preprint UR-1413, March 1995; hep-th/9503032.

[108] T. Haugset and F. Ravndal, "Scale Anomalies in Nonrelativistic Field Theories in 2 + 1 Dimensions", *Phys. Rev. D* **49** (1994) 4299.

[109] N. Hitchin, "The Self-Duality Equations on a Riemann Surface", *Proc. Lond. Math. Soc.* **55** (1987) 59.

[110] Z. Hlousek and D. Spector, "Supersymmetric Anyons", *Nucl. Phys. B* **344** (1990) 763.

[111] Z. Hlousek and D. Spector, "Why Topological Charges Imply Extended Supersymmetry", *Nucl. Phys. B* **370** (1992) 143.

[112] Z. Hlousek and D. Spector, "Bogomol'nyi Explained", *Nucl. Phys. B* **397** (1993) 173.

[113] J. Hong, Y. Kim and P-Y. Pac, "Multivortex Solutions of the Abelian Chern-Simons-Higgs Theory", *Phys. Rev. Lett.* **64** (1990) 2330.

[114] J. Hoppe, *Lectures on Integrable Systems*, Lecture Notes in Physics Vol. m 10 (Springer, Berlin 1992).

[115] M. Hotta, "Imported Symmetry of Two Breathing Modes in Chern-Simons Theory With External Magnetic Field", *Prog. Theor. Phys.* **86** (1991) 1289.

[116] L. Hua and C. Chou, "Dynamics of Non-Relativistic Chern-Simons Solitons", *Phys. Lett. B* **308** (1993) 286.

[117] J. Humphreys, *Introduction to Lie Algebras and Representation Theory* (Springer-Verlag 1990).

[118] R. Iengo and K. Lechner, "Anyon Quantum Mechanics and Chern-Simons Theory", *Phys. Rep.* **213** (1992) 179.

[119] Y. Ipekoglu, M. Leblanc and M. T. Thomaz, "Thermal and Quantum Fluctuations in Supersymmetric Chern-Simons Theory", *Ann. Phys.* **214** (1992) 160.

[120] E. Ivanov, "Chern-Simons Matter Systems with Manifest N=2 Supersymmetry", *Phys. Lett. B* **268** (1991) 203.

[121] R. Jackiw, "Introducing Scale Symmetry", *Phys. Today* **25** (1972) 23.

[122] R. Jackiw, "Quantum Meaning of Classical Field Theory", *Rev. Mod. Phys.* **49** (1977) 681.

[123] R. Jackiw, "Topics in Planar Physics", in *Physics, Geometry and Topolgy*, Proceedings of Banff NATO Summer Institute, 1989, H. C. Lee, Ed., NATO ASI Series B: Physics Vol. 238 (Plenum Press, New York, 1990).

[124] R. Jackiw, "Dynamical Symmetry of the Magnetic Vortex", *Ann. Phys.* **201** (1990) 83-116.

[125] R. Jackiw and S-Y. Pi, "Soliton Solutions to the Gauged Nonlinear Schrödinger Equation on the Plane", *Phys. Rev. Lett.* **64** (1990) 2969.

[126] R. Jackiw and S-Y. Pi, "Classical and Quantum Nonrelativistic Chern-Simons Theory", *Phys. Rev. D* **42** (1990) 3500.

[127] R. Jackiw and E. Weinberg, "Self-Dual Chern-Simons Vortices", *Phys. Rev. Lett.* **64** (1990) 2334.

[128] R. Jackiw, S-Y. Pi and E. Weinberg, "Topological and Nontopological Solitons in Relativistic and Nonrelativistic Chern-Simons Theory", talk at Boston PASCOS 1990:573-588 (QCD161: I69:1990).

[129] R. Jackiw, K. Lee and E. Weinberg, "Self-Dual Chern-Simons Solitons", *Phys. Rev. D* **42** (1990) 3488.

[130] R. Jackiw and S-Y. Pi, "Reply to: Comment on 'Soliton Solutions to the Gauged Nonlinear Schrödinger Equation on the Plane"', *Phys. Rev. Lett.* **66** (1991) 2682.

[131] R. Jackiw and S-Y. Pi, "Time-Dependent Chern-Simons Solitons and Their Quantization", *Phys. Rev. D* **44** (1991) 2524.

[132] R. Jackiw and S-Y. Pi, "Semiclassical Landau Levels of Anyons", *Phys. Rev. Lett.* **67** (1991) 415.

[133] R. Jackiw and S-Y. Pi, "Self-Dual Chern-Simons Solitons", *Prog. Theor. Phys. Suppl.* **107** (1992) 1.

[134] R. Jackiw and S-Y. Pi, "Finite and Infinite Symmetries in $2 + 1$ Dimensional Field Theory", *Nucl. Phys. B (Proc. Suppl.)* **33C** (1993) 104-113.

[135] R. Jackiw, "Delta-Function Potentials in Two- and Three-Dimensional Quantum Mechanics", in *M. Beg Memorial Volume*, A. Ali and P. Hoodbhoy, Eds. (World Scientific, Singapore 1991).

[136] L. Jacobs, A. Khare, C. Kumar and S. Paul, "The Interaction of Chern-Simons Vortices", *Int. J. Mod. Phys. A* **6** (1991) 3441.

[137] A. Jaffe and C. Taubes, *Vortices and Monopoles* (Birkhäuser 1980).

[138] D. Jatkar and A. Khare, "Peculiar Charged Vortices in Higgs Model with Pure Chern-Simons Term", *Phys. Lett. B* **236** (1990) 283.

[139] D. Kabat, "Canonical Quantization of Abelian Chern-Simons Solitons", *Phys. Lett. B* **281** (1992) 265-270.

[140] H-C. Kao and K. Lee, "Self-Dual $SU(3)$ Chern-Simons Higgs Systems", *Phys. Rev. D* **50** (1994) 6626-6632.

[141] H-C. Kao and K. Lee, "Self-Dual Chern-Simons Higgs Systems with an N=3 Extended Supersymmetry", *Phys. Rev. D* **46** (1992) 4691.

[142] H-C. Kao, "Self-Dual Yang-Mills Chern-Simons Higgs Systems with an N=3 Extended Supersymmetry", *Phys. Rev. D* **50** (1994) 2881.

[143] H-C. Kao, K. Lee, C. Lee and T. Lee, "The Chern-Simons Coefficient in the Higgs Phase", *Phys. Lett. B* **341** (1994) 181.

[144] H-C. Kao, "Generalizing the Coleman-Hill Theorem", preprint June 1995; hep-th/9506093.

[145] H-C. Kao, K. Lee and T. Lee, "The Chern-Simons Coefficient in Supersymmetric Yang-Mills Chern-Simons Theories", preprint June 1995; hep-th/9506170.

[146] A. Khare, "Rigorous Lower Bound on the Flux of Nontopological Self-Dual Chern-Simons Vortices", *Phys. Lett. B* **263** (1991) 227.

[147] A. Khare, R. MacKenzie, P. Panigrahi and M. Paranjape, "Spontaneous Symmetry Breaking and the Renormalization of the Chern-Simons Term", Montréal preprint UdeM-LPS-TH-93-150; hep-th/9306027.

[148] A. Khare, R. MacKenzie and M. Paranjape, "On the Coleman-Hill Theorem", *Phys. Lett. B* **343** (1995) 239.

[149] C. Kim, C. Lee, P. Ko, B.-H. Lee and H. Min, "Schrödinger Fields on the Plane with $[U(1)]^N$ Chern-Simons Interactions and Generalized Self-Dual Solitons", *Phys. Rev. D* **48** (1993) 1821.

[150] S-K. Kim, K-S. Soh and J-H. Yee, "Index Theory for the Nonrelativistic Chern-Simons Solitons", *Phys. Rev. D* **42** (1990) 4139.

[151] S-K. Kim, K-S. Soh and J-H. Yee, "Inversion Symmetry and Flux Quantization in the Nonrelativistic Chern-Simons Solitons", *Phys. Rev. D* **46** (1992) 1882.

[152] S. Kim and H. Min, "Statistical Interactions between Chern-Simons Vortices", *Phys. Lett. B* **281** (1992) 81.

[153] S-J. Kim, "Absence of Scale Anomaly to all Orders in Nonrelativistic Self-Dual Chern-Simons Theory", *Phys. Lett. B* **343** (1995) 244.

[154] W. Kim and C. Lee, "Schrödinger Fields on the Plane with Nonabelian Chern-Simons Interactions", *Phys. Rev. D* **49** (1994) 6829.

[155] Y. Kim and K. Lee, "Vortex Dynamics in Self-Dual Chern-Simons-Higgs Systems", *Phys. Rev. D* **49** (1994) 2041.

[156] M. Knecht, R. Pasquier and J. Y. Pasquier, "Painlevé Analysis and Integrability Properties of a 2 + 1 Nonrelativistic Field Theory", preprint IPNO-TH-95-05, Jan 1995; hep-th/9502128.

[157] B. Kostant, "The Principal 3-Dimensional Subgroup and the Betti Numbers of a Complex Simple Lie Group", *Amer. J. Math.* **81** (1959) 973.

[158] B. Kostant, "The Solution to a Generalized Toda Lattice and Representation Theory", *Adv. Math.* **34** (1979) 195.

[159] C. Kumar and A. Khare, "Charged Vortex of Finite Energy in Non-abelian Gauge Theories with Chern-Simons Term", *Phys. Lett. B* **178** (1986) 395.

[160] A. Kupiainen and J. Mickelsson, "What is the Effective Action in Two Dimensions?", *Phys. Lett. B* **185** (1987) 107.

[161] M. Leblanc, G. Lozano and H. Min, "Extended Superconformal Galilean Symmetry in Chern-Simons Matter Systems", *Ann. Phys.* **219** (1992) 328.

[162] M. Leblanc and M. T. Thomaz, "Maxwell-Chern-Simons Theory and an Ambiguity in Chern-Simons Perturbation Theory", *Phys. Lett. B* **281** (1992) 259.

206

[163] B.-H. Lee, C. Lee and H. Min, "Supersymmetric Chern-Simons Vortex Systems and Fermion Zero Modes", *Phys. Rev. D* **45** (1992) 4588.

[164] B.-H. Lee and H. Min, "Quantum Aspects of Supersymmetric Maxwell-Chern-Simons Solitons", *Phys. Rev. D* **51** (1995) 4458.

[165] C. Lee, K. Lee and E. Weinberg, "Supersymmetry and Self-Dual Chern-Simons Systems", *Phys. Lett. B* **243** (1990) 105.

[166] C. Lee, K. Lee, H. Min, "Self-Dual Maxwell-Chern-Simons Solitons", *Phys. Lett. B* **252** (1990) 79.

[167] C. Lee, H. Min and C. Rim, "Zero Modes of the Selfdual Maxwell Chern-Simons Solitons", *Phys. Rev. D* **43** (1991) 4100.

[168] C. Lee, "Instantons, Monopoles and Vortices", Lectures at 13th Symposium on Theoretical Physics, *Field Theory and Mathematical Physics*, Mt. Sorak, Korea, June-July 1994, published in the Proceedings, J. E. Kim, Ed. (Mineumsa, Seoul, 1995).

[169] K. Lee, "Relativistic nonabelian self-dual Chern-Simons systems ", *Phys. Lett. B* **255** (1991) 381.

[170] K. Lee, "Self-Dual Nonabelian Chern-Simons Solitons", *Phys. Rev. Lett.* **66** (1991) 553.

[171] K. Lee and P. Yi, "Self-Dual Anyons in Uniform Background Fields", Columbia preprint CU-TP-668; hep-th/9501043.

207

[172] T. Lee and H. Min, "Bogomol'nyi Equations for Solitons in Maxwell-Chern-Simons Gauge Theories with Magnetic Moment Interaction Term", *Phys. Rev. D* **50** (1994) 7738.

[173] T. Lee and P. Oh, "Coherent State Quantization of $SU(N)$ Non-abelian Chern-Simons Particles", *Phys. Lett. B* **319** (1994) 497.

[174] T. Lee and P. Oh, "Nonabelian Chern-Simons Quantum Mechanics and Nonabelian Aharonov-Bohm Effect", *Ann. Phys.* **235** (1994) 413.

[175] A. Lerda, *Anyons : Quantum Mechanics of Particles with Fractional Statistics*, Lecture Notes in Physics Vol. m 14 (Springer, Berlin 1992).

[176] D. Levi, L. Vinet and P. Winternitz, "Symmetries and Conditional Symmetries of a Nonrelativistic Chern-Simons System", *Ann. Phys.* **230** (1994) 101.

[177] A. Leznov and M. Saveliev, "Representation of Zero Curvature for the System of Nonlinear Partial Differential Equations $x_{\alpha,z\bar{z}} = exp(kx)_\alpha$ and its Integrability", *Lett. Math. Phys.* **3** (1979) 389.

[178] A. Leznov, "On the Complete Integrability of a Nonlinear System of Partial Differential Equations in Two Dimensional Space", *Theor. Math. Phys.* **42** (1980) 225.

[179] A. Leznov and M. Saveliev, "Representation Theory and Integration of Nonlinear Spherically Symmetric Equations of Gauge Theories", *Comm. Math. Phys.* **74** (1980) 111.

[180] A. Leznov, M. Saveliev and V. Smirnov, "Theory of Group Representations and Integration of Nonlinear Dynamical Systems", *Theor. Math. Phys.* **48** (1981) 565.

[181] A. Leznov and V. Smirnov, "Graded Algebras of the Second Rank and Integration of Nonlinear Equations", *Lett. Math. Phys.* **5** (1981) 31.

[182] J. Liouville, "Sur l'équation aux différences partielles $\frac{d^2}{dudv}\log\lambda \pm \frac{\lambda}{2a^2} = 0$", *Journ. Math. Pures Appl.* **18** (1853) 71.

[183] Q. Liu, "Chern-Simons Soliton Dynamics in Modular Parameter Space", *Phys. Lett. B* **321** (1994) 219-222.

[184] G. Lozano, "Ground State Energy for Nonrelativistic Bosons Coupled to Chern-Simons Gauge Fields", *Phys. Lett. B* **283** (1992) 70.

[185] J. Lykken, J. Sonnenschein and N. Weiss, "The Theory of Anyonic Superconductivity", *Int. J. Mod. Phys. A* **6** (1991) 5155.

[186] N. Manton, "A Remark on the Scattering of BPS Monopoles", *Phys. Lett. B* **110** (1982) 54.

[187] C. Manuel and R. Tarrach, "Contact Interaction of Anyons", *Phys. Lett. B* **268** (1991) 222.

209

[188] C. Manuel and R. Tarrach, "Contact Interactions and Dirac Anyons", *Phys. Lett. B* **301** (1993) 72.

[189] C. Manuel and R. Tarrach, "Perturbative Renormalizations in Quantum Mechanics", *Phys. Lett. B* **328** (1994) 113.

[190] L. Martina, O. Pashaev and G. Soliani, "Self-Dual Chern-Simons Solitons in Nonlinear σ-Models", *Mod. Phys. Lett. A* **8** (1993) 3241.

[191] J. McCabe and S. Ouvry, "Perturbative Three Body Spectrum and the Third Virial Coefficient in the Anyon Model", *Phys. Lett. B* **260** (1991) 113.

[192] A. Mikhailov, M. Olshanetsky and A. Perelomov, "Two Dimensional Generalized Toda Lattice", *Comm. Math. Phys.* **79** (1981) 473.

[193] C. Montonen and D. Olive, "Magnetic Monopoles as Gauge Particles", *Phys. Lett. B* **72** (1977) 117.

[194] C. Montonen, "The Many-Anyon Problem", Lectures at the VI Mexican School of Particles and Fields, Villahermosa, Tabasco, October 1994; Helsinki preprint HU-TFT-95-12; hep-th/9502071.

[195] B. Nagel, "Comment on the Born Approximation in Aharonov-Bohm Scattering", *Phys. Rev. D* **32** (1985) 3328.

[196] H. Nielsen and P. Olesen, "Vortex-line models for Dual Strings", *Nucl. Phys.* **B61** (1973) 45.

[197] H. Nishino and S. J. Gates, "Chern-Simons Theories with Super-symmetries in Three Dimensions ", *Int. J. Mod. Phys. A* **8** (1993) 3371.

[198] P. Olesen, "Soliton Condensation in some Self-Dual Chern-Simons Theories", *Phys. Lett. B* **265** (1991) 361, **(E) 267** (1991) 541.

[199] O. Pashaev, "Integrable Chern-Simons Gauge Field Theory in $2+1$ Dimensions", Trieste preprint (April 1995); hep-th/9505178.

[200] S. Paul and A. Khare, "Self-Dual Factorization of the Proca Equation with Chern-Simons Term in $4k-1$ Dimensions", *Phys. Lett. B* **171** (1986) 244.

[201] S. Paul and A. Khare, "Charged Vortices in an Abelian Higgs Model with Chern-Simons Term", *Phys. Lett. B* **174** (1986) 420, (E) *B* **177** (1986) 453.

[202] A. Perelomov, *Generalized Coherent States and their Applications*, (Springer-Verlag, New York) 1985.

[203] B. Piette and W. Zakrzewski, "General Solutions of the $U(3)$ and $U(4)$ Chiral Sigma Models in Two Dimensions", *Nucl. Phys.* **B300** (1988) 207.

[204] B. Piette and W. Zakrzewski, "Some Classes of General Solutions

of the $U(N)$ Chiral σ Models in Two Dimensions", *J. Math. Phys.* **30** (1989) 2233.

[205] R. Pisarski and S. Rao, "Topologically Massive Chromodynamics in the Perturbative Regime", *Phys. Rev. D* **32** (1985) 2081.

[206] K. Pohlmeyer, "Integrable Hamiltonian Systems and Interactions Through Constraints", *Comm. Math. Phys.* **46** (1976) 207.

[207] A. Polyakov and P. Wiegmann, "Theory of Nonabelian Goldstone Bosons in Two Dimensions", *Phys. Lett. B* **131** (1983) 121.

[208] A. Polychronakos, "Abelian Chern-Simons Theories in 2+1 Dimensions", *Ann. Phys.* **203** (1990) 231.

[209] R. Prange and S. Girvin, *The Quantum Hall Effect*, (Springer-Verlag, New York, 1990).

[210] M. Prasad and C. Sommerfield, "Exact Classical Solution for the 't Hooft Monopole and the Julia-Zee Dyon", *Phys. Rev. Lett.* **35** (1975) 760.

[211] R. Rajaraman, *Solitons and Instantons* (North-Holland 1982).

[212] C. Rebbi and G. Soliani, *Solitons and Particles* (World Scientific 1984).

[213] M. Reuter, "The Maslov Index in Chern-Simons Quantum Mechanics", *Phys. Rev. D* **42** (1990) 2763.

[214] P. Ruback, "Vortex String Motion in the Abelian Higgs Model", *Nucl. Phys.* **B296** (1988) 669.

[215] S. Ruijsenaars, "The Aharonov-Bohm Effect and Scattering Theory", *Ann. Phys.* **146** (1983) 1.

[216] T. Samols, "Hermiticity of the Metric on Vortex Moduli Space", *Phys. Lett. B* **244** (1990) 285.

[217] T. Samols, "Vortex Scattering", *Comm. Math. Phys.* **145** (1992) 149.

[218] J. Schiff, "Integrability of Chern-Simons-Higgs and Abelian Higgs Vortex Equations in a Background Metric", *J. Math. Phys.* **32** (1991) 753.

[219] J. Schonfeld, "A Mass Term for Three-Dimensional Gauge Fields", *Nucl. Phys. B* **185** (1981) 157.

[220] J. Schwinger, "Magnetic Charge and Quantum Field Theory", *Phys. Rev.* **144** (1966) 1087.

[221] J. Schwinger, "Electric- and Magnetic-Charge Renormalization: I", *Phys. Rev.* **151** (1966) 1048.

[222] J. Schwinger, "Electric- and Magnetic-Charge Renormalization: II", *Phys. Rev.* **151** (1966) 1055.

[223] G. Semenoff, "Chern-Simons Gauge Theory and Spin Statistics

Connection in Two Dimensional Quantum Mechanics", in *Physics, Geometry and Topolgy*, Proceedings of Banff NATO Summer Institute, 1989, H. C. Lee, Ed., NATO ASI Series B: Physics Vol. 238 (Plenum Press, New York, 1990).

[224] W. Siegel, "Unextended Superfields in Extended Supersymmetry", *Nucl. Phys. B* **156** (1979) 135.

[225] J. Spruck and Y. Yang, "The Existence of Nontopological Solitons in the Self-Dual Chern-Simons Theory", *Comm. Math. Phys.* **149** (1992) 361.

[226] J. Spruck and Y. Yang, "Topological Solutions in the Self-Dual Chern-Simons Theory: Existence and Approximation", *Ann. de l'Inst. H. P. - Anal. Non Lin.* **12** (1995) 75.

[227] J. Spruck and Y. Yang, "Existence Theorems for Periodic Nonrelativistic Maxwell-Chern-Simons Solitons", preprint 1994.

[228] P. Srivastava and K. Tanaka, "On the Self-Duality Condition in Chern-Simons Systems", *Phys. Lett. B* **256** (1991) 427.

[229] M. Stone, *Quantum Hall Effect*, (World Scientific, Singapore, 1992).

[230] M. Toda, "Studies of a Nonlinear Lattice", *Phys. Rep.* **8** (1975) 1.

[231] M. Torres, "Bogomol'nyi Limit for Nontopological Solitons in a Chern-Simons Model with Anomalous Magnetic Moment", *Phys. Rev. D* **46** (1992) 2295.

[232] C. A. Trugenberger, "The Anyon Fluid in the Bogoliubov Approximation", *Phys. Rev. D* **45** (1992) 3807.

[233] C. A. Trugenberger, "Ground State and Collective Excitations of Extended Anyons", *Phys. Lett. B* **288** (1992) 121.

[234] C. A. Trugenberger, "Topics in Planar Gauge Theories", Lectures given at the *3ème Cycle de la Physique en Suisse Romande*, Université de Lausanne, 1994/95.

[235] K. Uhlenbeck, "Harmonic Maps into Lie Groups (Classical Solutions of the Chiral Model)", preprint (1985), *J. Diff. Geom.* **30** (1989) 1.

[236] M. Valle Basagoiti, "Pressure in Chern-Simons Field Theory to Three Loop Order", *Phys. Lett. B* **306** (1993) 307.

[237] G. Valli, "On the Energy Spectrum of Harmonic Two-Spheres in Unitary Groups", *Topology* **27** (1988) 129.

[238] P. Valtancoli, "Classical Chern-Simons Vortices on Curved Space", *Int. J. Mod. Phys. A* **7** (1992) 4335.

[239] E. Verlinde, "A Note on Braid Statistics and the Nonabelian Aharonov-Bohm Effect", in Proceedings of *Modern Quantum Field Theory*, Bombay, India, Jan 8-14 (1990), S. Das et al, Eds.

[240] R. Wang, "The Existence of Chern-Simons Vortices", *Comm. Math. Phys.* **137** (1991) 587.

[241] R. Ward, "Integrable and Solvable Systems and Relations Among Them", *Phil. Trans. Roy. Soc. Lond.* **A315** (1985) 451.

[242] R. Ward, "Multidimensional Integrable Systems", in *Field Theory, Quantum Gravity and Strings II*, H. de Vega and N. Sanchez (Eds.) (Springer Lecture Notes in Physics #280, 1987).

[243] R. Ward, "Integrable Systems in Twistor Theory", in *Twistors in Mathematics and Physics*, eds. T. Bailey and R. Baston.

[244] R. Ward, "Classical solutions of the Chiral Model, Unitons and Holomorphic Vector Bundles", *Comm. Math. Phys.* **128** (1990) 319.

[245] E. Weinberg, "Multivortex Solutions of the Landau-Ginzburg Equations", *Phys. Rev. D* **19** (1979) 3008.

[246] X. G. Wen and A. Zee, "On the Possibility of a Statistics Changing Phase Transition", *J. Phys. France* **50** (1989) 1623.

[247] F. Wilczek, *Fractional Statistics and Anyonic Superconductivity*, (World Scientific, Singapore, 1990).

[248] E. Witten and D. Olive, "Supersymmetry Algebras that Include Topological Charges", *Phys. Lett. B* **78** (1978) 97.

[249] E. Witten, "Quantum Field Theory and the Jones Polynomial", *Comm. Math. Phys.* **121** (1989) 351.

[250] S. K. Wong, "Field and Particle Equations for the Classical Yang-Mills Field and Particles with Isotopic Spin", *Nuovo Cim.* **65A** (1970) 689.

[251] J. C. Wood, "Explicit Construction and Parametrization of Harmonic Two-Spheres in the Unitary Group", *Proc. Lond. Math. Soc.* **58** (1989) 608.

[252] C. N. Yang, "Condition of Self-Duality for $SU(2)$ Gauge Fields on Euclidean Four-Dimensional Space", *Phys. Rev. Lett.* **38** (1977) 1377.

[253] Y. Yang, "A Generalized Self-Dual Chern-Simons Higgs Theory", *Lett. Math. Phys.* **23** (1991) 179.

[254] Y. Yoon, "Zero Modes of the Nonrelativistic Self-Dual Chern-Simons Vortices on the Toda Backgrounds", *Ann. Phys.* **211** (1991) 316.

[255] V. Zakharov (Ed.), *What is Integrability?*, Springer Series in Nonlinear Science (Springer 1991).

[256] W. Zakrzewski, *Low Dimensional Sigma Models* (Adam Hilger 1989).

[257] A. Zee, "Long-Distance Physics of Topological Fluids", *Prog. Theor. Phys. Suppl.* **107** (1992) 77.

Springer-Verlag
and the Environment

We at Springer-Verlag firmly believe that an international science publisher has a special obligation to the environment, and our corporate policies consistently reflect this conviction.

We also expect our business partners – paper mills, printers, packaging manufacturers, etc. – to commit themselves to using environmentally friendly materials and production processes.

The paper in this book is made from low- or no-chlorine pulp and is acid free, in conformance with international standards for paper permanency.

Lecture Notes in Physics

For information about Vols. 1–425
please contact your bookseller or Springer-Verlag

New Series m: Monographs